Doctors, Honour and the Law

Also by Andreas-Holger Maehle

DRUGS ON TRIAL: Experimental Pharmacology and Therapeutic Innovation in the Eighteenth Century

HISTORICAL AND PHILOSOPHICAL PERSPECTIVES ON BIOMEDICAL ETHICS: From Paternalism to Autonomy? (Co-edited with Johanna Geyer-Kordesch)

JOHANN JAKOB WEPFER (1620–1695) ALS TOXIKOLOGE

KRITIK UND VERTEIDIGUNG DES TIERVERSUCHS: DIE ANFÄNGE DER DISKUSSION IM 17. UND 18. JAHRHUNDERT

Doctors, Honour and the Law

Medical Ethics in Imperial Germany

Andreas-Holger Maehle
Professor of History of Medicine and
Medical Ethics, Durham University

First published 2009 by
PALGRAVE MACMILLAN

Palgrave Macmillan in the UK is an imprint of Macmillan Publishers Limited, registered in England, company number 785998, of Houndmills, Basingstoke, Hampshire RG21 6XS.

Palgrave Macmillan in the US is a division of St Martin's Press LLC, 175 Fifth Avenue, New York, NY 10010.

Palgrave Macmillan is the global academic imprint of the above companies and has companies and representatives throughout the world.

Palgrave® and Macmillan® are registered trademarks in the United States, the United Kingdom, Europe and other countries.

ISBN-13: 978-0-230-55330-9 hardback
ISBN-10: 0-230-55330-3 hardback

This book is printed on paper suitable for recycling and made from fully managed and sustained forest sources. Logging, pulping and manufacturing processes are expected to conform to the environmental regulations of the country of origin.

A catalogue record for this book is available from the British Library.

Library of Congress Cataloging-in-Publication Data

Maehle, Andreas-Holger, 1957–
    Doctors, honour, and the law : medical ethics in imperial Germany / Andreas-Holger Maehle.
        p. cm.
    Includes bibliographical references.
    ISBN 978-0-230-55330-9
        1. Medical ethics – Germany – History – 19th century. 2. Physician and patient – Germany – History – 19th century. 3. Courts of honor – Germany – History – 19th century. I. Title.

R724.M1626 2009
174.20943'09034—dc22   2008050871

10 9 8 7 6 5 4 3 2 1
18 17 16 15 14 13 12 11 10 09

Printed and bound in Great Britain by
CPI Antony Rowe, Chippenham and Eastbourne

Contents

Tables

Acknowledgements

My research on medical ethics in Imperial Germany started in early 1994, when I took up a Wellcome University Award in History of Medicine in the Philosophy Department of Durham University. Nearly 15 years later, and after an intensive finale during a Wellcome Research Leave Award in 2007/08, I am pleased to be able to give this book to the world and to thank all those who have helped me along the way.

Naturally, my first thanks go to the Wellcome Trust, which has generously supported this project. Additional funding was kindly provided by the British Council and by Durham University's Research Committee.

In the course of my research I have benefited from discussions with many colleagues as well as with public audiences at lectures, in research seminars and at conferences, in London, Durham, Aberdeen, Freiburg (Breisgau), Göttingen, Bonn, Lucerne, Oslo, Halifax (Nova Scotia) and Rochester (New York). I would like to thank the many discussants who have shaped my thinking and writing at these occasions. Moreover, I am indebted to the Wellcome Trust's anonymous referees of this book project for their useful suggestions. In this regard special thanks also go to my colleagues in Durham's Centre for the History of Medicine and Disease in the Wolfson Research Institute, Lutz Sauerteig and Sebastian Pranghofer, the latter also for his efficient research assistance. I am furthermore grateful for the advice that I received at the state archives of Berlin, Dresden and Munich. Access to the published sources was provided with friendly professionalism by the university libraries of Durham, Freiburg, Berlin, Göttingen, Munich and Hamburg, the Max Planck Institute for Foreign and International Criminal Law in Freiburg, the Wellcome Library and the British Library.

Our Centre administrator at Durham, Katherine Smith, kindly took over the task of checking the text for style before it went to the publisher. Helpful for the final revisions of the book were the comments by Palgrave Macmillan's anonymous reviewer. Thanks to

Michael Strang and Ruth Ireland at Palgrave Macmillan for their support in the production of this book.

Last but not the least, I would like to thank my wife Jill for some stimulating comments on the topic from a lawyer's point of view, and our kids Rupert and Caroline-Sophie, for putting up with their dad's obsession with the history of medicine.

Introduction

In 1902, the Berlin psychiatrist Albert Moll (1862–1939) introduced his 650-page handbook of medical ethics, *Ärztliche Ethik*, with the following critical observation:

> On various occasions, in assemblies of doctors, in journal articles, and also in independent monographs, the importance of doctors' ethics has been discussed in recent years. It has been proposed several times to teach medical ethics at the universities... From another perspective, it has been recommended to publish a *Codex deontologicus* as a kind of medical statute book. In these discussions deontology was taken to be more or less synonymous with the doctrine of ethical duties, but mostly... the so-called professional duties were overemphasised, even seen as identical with ethical duties. This mistake also becomes apparent in many medical circles, where offences against professional duties are judged very harshly, but censure for a serious violation of ethical duties is used very sparingly.[1]

Why had medical ethics become so important for doctors towards the end of the nineteenth century? What kind of medical ethics was criticised by Moll, and how did it come about that doctors seemed to pay more attention to duties towards their profession than to ethical obligations vis-à-vis their patients? What were those professional duties, and why and how was adherence to them monitored so closely? If a code of medical duties was envisaged to be as binding as a statute book, what was the relationship between medical ethics

1

and the law? These and related questions will be addressed in this book, which for the first time provides a comprehensive perspective on doctors' professional ethics in Germany from the foundation of the Second Reich in 1871 to the start of the First World War.[2]

Some specific aspects of this topic have been addressed in the earlier historiography. For example, Paul Weindling has traced the affinities of German medicine and public health policy to Social Darwinist and eugenic ideas from national unification to the Nazi period. As Weindling concludes, the 'medical profession's quest for social power was pursued through the channels of eugenics'.[3] Another milestone in the characterisation of doctors' ethics in Imperial Germany is Barbara Elkeles' study of the practices and discussion of human experimentation during that period. It highlights the disregard of contemporary medical researchers for the consent and safety of their human subjects, which stemmed from a fanatical belief in medical progress through science and was facilitated by an authoritarian culture in the public hospitals and university clinics.[4]

Remarkably, however, the *professional ethics as such* of doctors in Imperial Germany has only been brought into focus in recent years. While an overview of medical–deontological writings in nineteenth-century Germany was given by Ulrich Brand as early as in 1977, a deeper engagement with this subject matter has occurred only after groundbreaking social historical research on medical professionalisation had been carried out.[5] As far as Germany is concerned, the study by Claudia Huerkamp of the social rise of doctors in Prussia during the nineteenth century has been particularly important.[6] It has been helpfully complemented by a similar study on Wurttemberg by Annette Drees; a broader history of the politics of the German medical profession between 1800 and 1996 by Robert Jütte and collaborators and a collection of essays on German medicine and public health in the nineteenth and twentieth centuries edited by Manfred Berg and Geoffrey Cocks.[7]

As this research has made clear, the professionalisation of German doctors was much more closely linked to state policies than that of their British and American colleagues.[8] Chancellor Otto von Bismarck's introduction of compulsory health insurance for workers in 1883 had, in the long run, a huge impact on the development of the German medical profession, including its notions of professional duties and ethics. On the one hand, the health insurance system brought more patients, especially from the lower social classes, into doctors' practices. The proportion of the population who were state-insured rose from approximately 9 per cent in 1885 to 23 per cent in 1914. If one includes family members who were covered by the bread-winner's insurance and people with a non-state insurance, about one-third of the population were compulsorily insured by the start of the First World War.[9] On the other hand, the increasing import-ance of insured patients led to fierce competition among doctors for contracts with the insurance funds, especially as the number of doc-tors rose significantly during the 1880s and 1890s. In the German Reich, there were approximately 3,200 inhabitants per doctor in 1885. By 1911, the number of inhabitants per doctor was approxi-mately 2,100.[10]

The relationship of the medical profession to the health insurance organisations was fraught with conflict, as the latter controlled doc-tors' fees and limited the number of doctors who were admitted to panel practice. Only with the foundation of the *Hartmann-Bund*, the doctors' association for the protection of their economic interests, in 1900, did the medical profession start to gain power in the struggle with the health insurance boards, as it then had a body that negoti-ated collective contracts and exercised pressure by organising doc-tors' strikes. Medical professional ethics in Imperial Germany was as much about defusing competition among doctors as about enforcing solidarity vis-à-vis the health insurance boards. Relative 'overcrowd-ing' of the profession and the increasingly powerful insurance sys-tem were prime movers in medical organisation, particularly in the formation of disciplinary tribunals or 'medical courts of honour', as they were called. The first chapter of this book therefore discusses the development of these courts of honour, their activities and the kind of ethics that they promulgated.[11] The disciplinary tribunals were, however, not merely instruments of professional politics. Through their judgements they articulated and implemented con-temporary codes of honour. The notion of professional honour of medical men – women were only gradually admitted to medical studies in the first decade of the twentieth century – thus forms another focus of this book.[12]

Discussions about medical ethics in Imperial Germany were also strongly influenced by its often-controversial relationship to the law. In particular, two sections of the Reich Penal Code of 1871, section 223 on physical injury and section 300 on professional secrecy, were highly relevant for doctors. The second and third chapters of this book examine those important discussions. As far as the issue of medical confidentiality was concerned, the rising influence of public health made itself felt through increasing calls for notification of infectious, especially venereal diseases. Individual interest in privacy was balanced against public interest in protection against contagious diseases and epidemics.[13] Doctors as well as the legal profession contributed extensively to this debate. Lawyers and medics were also the main contributors to contemporary discussions about whether surgery without information and explicit consent of the patient had to be seen (and punished) as physical injury. While doctors uniformly rejected such a legal interpretation, lawyers were divided in their opinions on this issue. Some acknowledged therapeutic intention as a sufficient reason to treat without patient consent, but others insisted on the self-determination of the individual. Such self-determination had come under threat with the significant expansion of surgical procedures that occurred after the introduction of anaesthesia in the 1840s and of antisepsis in the late 1860s. Surgeons felt increasingly confident about the success of their treatments, so that patients' personal wishes or views mattered little to them.[14] This paternalistic mentality may also have been nurtured by the introduction of compulsory vaccination against smallpox in 1874. Although such legal compulsion generated considerable anti-vaccinationist protests, it meant that a medical intervention into people's bodies had been condoned for the first time by the authority of the state – in the name of (public) health.[15]

In this climate of professional, legal and scientific changes for medicine, doctors reflected and published extensively about their ethical duties and role in society. The fourth chapter of this book therefore examines the changing patterns of the German medical ethics literature from the 1870s to the First World War. The social challenges that doctors met through 'overcrowding' of their profession and the health insurance system were met by predominantly conservative responses which sought to preserve an idealised individual relationship between doctor and patient.[16] Paternalistic attitudes predominated among the medical authors of the deontological literature. They became manifest in issues such as truth-telling at the sickbed, palliative care of the dying and the treatment of women who had made an illegal abortion attempt. Only transitory efforts were made to enhance the patient's self-determination, before the tone of doctors' ethical writing turned increasingly authoritarian after the First World War.

Medical ethics in Imperial Germany was thus deeply entangled in professional, legal and social issues. This book is about understanding and elucidating this entanglement. It will show how doctors' ethical decision-making during the *Kaiserreich* was less guided by concern for patients' interests than by notions of male honour and professional reputation and by considerations of professional politics. Moreover, it will illustrate how medical men adhered to a paternalistic conception of the doctor–patient relationship, although they experienced increasing pressures from lawyers and patients themselves to recognise a right to self-determination of the sick individual. Following the traumatic experience of defeat in the First World War and against the background of the economic crises of the Weimar Republic, arguments for a priority of collective over individual interests gained ground, and doctors conceived their relationship to patients in increasingly authoritarian terms. Initiatives like that of Moll, who in 1902 had developed a detailed account of how medical ethics could be built upon a contract relationship between doctor and patient, were marginalised and forgotten. Instead, the image of the doctor as a ruler and leader, as disseminated by Erwin Liek (1878–1935) and other medical writers of the 1920s, became popular.

The study of doctors' ethics around 1900 thus involves issues that are relevant not only to the history of medicine as such but also to fields such as German history, biomedical ethics, legal history and cultural studies. Accordingly, this book has been written with a broad, interdisciplinary readership in mind.

1
Disciplining Doctors: Medical Courts of Honour and Professional Conduct

1.1 Introduction

Self-regulation, rather than regulation by the state, is a recognised criterion of a profession.[1] This applies also to the organisation of the medical profession in Imperial Germany, and yet, as this chapter will show, doctors sought the backing of the state in implementing the most tangible form of professional regulation: the so-called courts of honour or disciplinary tribunals for cases of misconduct.

The concept of courts of honour (*Ehrengerichte*) was initially linked to the practice of duelling. Although duelling had been banned in the Prussian General Common Law (1794) and in the Penal Codes for Prussia (1851), Bavaria (1861) and eventually for the whole German Reich (1871), it was still widely practised until the First World War, not only among aristocrats and military officers but also among civil servants, lawyers, doctors and other academics. Courts of honour for military officers had been formed in the early nineteenth century (e.g., in Prussia in 1808/1821; Saxony in 1833) as disciplinary tribunals and as institutions for arbitrating in cases of insults and violations of honour and thus, ideally, to avoid duelling.[2] Disciplinary proceedings against civil servants were regulated by Prussian law in 1844. Following this line, the first court of honour for lawyers was introduced by Royal decree in Prussia in 1847, and the legal profession generally adopted disciplinary tribunals with the Lawyers' Ordinance (*Rechtsanwaltsordnung*) for the German Reich in 1878.[3]

Based on the military and legal model, official courts of honour for medical doctors were introduced in several German states during the 1890s.[4] The following two sections of this chapter will delineate the process towards this institutionalised medical self-discipline with special reference to the developments in Prussia and Bavaria – the two largest states of Germany. The subsequent sections will then analyse and discuss the decisions of medical disciplinary tribunals and the kind of medical ethics that was promulgated by them. Crucial for understanding the decision-making of these courts is – as their name already indicates – the notion of professional honour around 1900, and I will therefore refer to Georg Simmel's contemporary conception of honour as a force for the cohesion of social groups. In addition, I will draw upon the ideas of sociologist Pierre Bourdieu on honour as a form of 'symbolic capital'.

1.2 Organisation of the German medical profession and the path to courts of honour for doctors in Prussia

The foundation of the German Empire in 1871 brought about important changes for the status and organisation of medical doctors. In 1869, a new Trade Ordinance (*Gewerbeordnung*) had been issued for the states of the North German Confederation, that is for Prussia and its allied states north of the River Main. Medical practice was included in this ordinance as one trade among many others. The new regulations established the principle of a so-called freedom to cure (*Kurierfreiheit*), which meant that anyone was permitted to practise medicine – only the use of the title *Arzt* (medical doctor) remained legally protected and restricted to fully qualified and licensed medical practitioners.[5] After the three German states south of the Main – Bavaria, Württemberg and Baden – had joined the Confederation with the proclamation of the German Empire, this Trade Ordinance became binding for all doctors in Germany from April 1871. Moreover, the political unification was reflected in the organisation of the medical profession. In 1873, the German Association of Doctors' Societies (*Deutscher Ärztevereinsbund*) was founded as a body that would represent the professional and economic interests of doctors at the national level. By 1874, almost 50 per cent of German doctors belonged to the Association through their membership of local medical societies, and by 1910 this figure had risen to over 80 per cent.[6]

The newly established status of medical practice as a 'free trade' had been a result, to a large extent, of lobbying by liberal sections of the medical profession, in particular the Berlin Medical Society (*Berliner Medizinische Gesellschaft*) during the 1860s. For a long time, they had resented the state control of doctors, which in principle had developed from the eighteenth-century concept of medical police, that is, the idea that healthcare should be centrally organised for the benefit of the population and the state.[7] Doctors in Prussia, for example, had not only to vow to practise conscientiously and to the best of their knowledge but were also required to swear an oath of allegiance to the King. Also, those doctors who did not hold a state office as town or district physician were obliged to submit regular reports on the health of the local population to the authorities. Doctors' fees had to comply with an official scale of charges. Most controversially, doctors were forbidden, under the Prussian Penal Code of 1851, to refuse medical assistance to anyone who requested it, day or night, and regardless of the patient's ability to pay.[8] A similar duty of assistance existed for Bavarian doctors on account of a police order of 1861. The Trade Ordinance of 1869/71 abolished this so-called obligation to cure (*Kurierzwang*) – a change that was initially welcomed by the medical profession, or a least by parts of it. According to the new regulations, doctors merely had the general duty of assistance in emergencies that applied to any citizen under section 360 of the 1871 Penal Code for the German Reich.[9]

However, from the mid-1870s resentment arose once again, as many doctors started to regret being reduced by law to the same level as common traders and non-licensed healers, a status that failed to reflect their bourgeois identity[10] and social ambitions. Doctors claimed that the Trade Ordinance had opened the floodgates to all kinds of unqualified lay healers, the so-called *Kurpfuscher*.[11] Also, heightened competition among licensed medical practitioners themselves within the new legal framework, and especially after Bismarck's introduction of compulsory health insurance for workers in 1883, led to fears that discipline in the profession might be weakened – to the detriment of doctors' reputation in the eyes of the public.[12] Looking towards another 'free profession', that of the then highly regarded lawyers, doctors strove to emulate their organisational model. The Lawyers' Ordinance of 1878 had organised the German lawyers through compulsory membership in state-authorised chambers, whose boards were granted disciplinary rights, that is they could act as courts of honour of first instance to deal with cases of alleged misconduct. Sanctions could range from warnings, reprimands and fines to exclusion from the legal profession. Appeals could be made to the Court of Honour (*Ehrengerichtshof*) for Lawyers, which was linked to the German Supreme Court (*Reichsgericht*) in Leipzig. The two senates of this court of honour had seven members each, four of whom had to be members of the Supreme Court.[13]

Following this model, doctors sought to be granted a state-authorised Doctors' Ordinance (*Ärzteordnung*), and the German Association of Doctors' Societies led the campaign for this objective. In February 1880, the chairman of the Association, Eduard Graf (1829–1895), presented a strategy paper entitled 'Aphorisms on Medical Reform' to its executive committee. The paper was subsequently sent to all member societies, as a basis for discussion at the Association's eighth annual meeting (*Ärztetag*) later that same year, and it was also published in the Association's journal, the *Ärztliches Vereinsblatt für Deutschland*.[14]

The title of Graf's paper had powerful resonances with the history of the medical profession. 'Aphorisms' were a reminder of the revered Hippocratic tradition (*Aphorisms* of Hippocrates). The term 'medical reform' (*Medizinalreform*) was linked to the efforts of German doctors in the revolution of 1848 to secure freedom from state control through self-organisation and a renewed campaign (in the 1860s) to achieve a national organisation of the profession and of public health.[15] In the immediate period after the failure of the revolution, activities of medical societies had temporarily slackened and tended to focus on scientific rather than political and social aims. Graf had been one of the doctors who revived their societies as organs of professional and social politics in the late 1860s.[16] As he stated in his 'Aphorisms', the medical profession should not jeopardise the achievements of the new Trade Ordinance, such as freedom in settling to open a medical practice, freedom to provide or refuse medical assistance and freedom for negotiation of fees. At the same time he regarded the following sentence as fundamental:

> The medical profession must not sink to the level of a mere trade, it has higher purposes, and duties to the community (and to the state), which it has to fulfil.[17]

The means for implementing this principle consisted, in his view, in strict professional organisation and in the subordination of the individual doctor under certain norms and laws. Voluntary organisation in medical societies had turned out to be insufficient in achieving these two aims, according to Graf. He therefore argued for the creation of state-recognised chambers of doctors (*Ärztekammern*). These chambers would advise the state in matters of medical legislation and public health, and they would seek to maintain the personal integrity of all members of the medical profession. This latter task required courts of honour linked to the chambers. The courts of first instance would arbitrate in quarrels between doctors, and judge and punish in cases of disciplinary offences. The more serious offences and punishments (up to withdrawal of a doctor's licence, the *Approbation*) would have to be confirmed by a higher disciplinary tribunal, which was also to serve as an appeal court. While the court of first instance was recommended to consist of elected medical doctors only, the court of second instance should also include a minority of state-appointed doctors and jurists.[18]

Debate on Graf's proposals quickly followed. For example, the medical society of the district of Koblenz held a meeting in May 1880, at which two doctors gave position papers, followed by resolutions of the assembly. These resolutions called for a reform of the Trade Ordinance that would outlaw medical practice by lay healers (the so-called *Kurpfuscher*) and abolish the use even of mere guidelines on doctor's fees. While the Koblenz society was prepared to agree to a Doctors' Ordinance that achieved these two aims, it explicitly did *not* support Graf's demand for state-authorised medical courts of honour.[19]

Nevertheless, Graf's 'Aphorisms' formed the blueprint for the organisational development of the German medical profession over the next 20 years. The Association's annual assembly of 1880 agreed that the practice of medical doctors should remain part of the Trade Ordinance for the time being, but that a specific Doctors' Ordinance should be formulated and official recognition be sought for it. This Doctors' Ordinance should eventually replace the relevant regulations of the Trade Ordinance. At the time, there were already some models for a tighter organisation of the medical profession. State-recognised representative bodies for doctors had been introduced in Baden (1864), Brunswick (1865), Saxony (1865), Bavaria (1871),

Wurttemberg (1875) and Hesse (1876). In Baden and Brunswick, these bodies also had some disciplinary powers.[20] However, the direct model for professional organisation at the national level and for the development of the system of courts of honour in Prussia was, as mentioned above, the Lawyers' Ordinance of 1878.

Using this ordinance as a template and following up Graf's proposals, the Karlsruhe Physician General, Adolf Hoffmann (1822–1899), provided in late 1880 a draft of a Doctors' Ordinance. In six sections the draft ordinance regulated (1) conditions for the medical practising licence (*Approbation*) and its possible withdrawal by the (general) courts (2) rights and duties of doctors (3) the formation and organisation of doctors' chambers in the various German states (4) central executive committees of these chambers (5) medical courts of honour (*ärztliche Ehrengerichte*) and (6) provisions for the transition from the Trade Ordinance to the Doctors' Ordinance. A key section (section 14) stated the obligation of doctors 'to practise in their profession conscientiously and to show through their conduct during professional practice as well as outside of it that they are worthy of the respect that their profession demands'.[21] This was a direct adaptation from section 28 of the Lawyers' Ordinance. It had two important implications: one was the proposal (in the same section 14) that doctors should agree upon professional codes that defined more closely the notion of appropriate conduct; and the other was the formation of medical courts of honour that would punish breaches of section 14 with warnings, reprimands, fines up to 500 Mark and temporal or permanent withdrawal of the right to vote in the membership elections of the relevant doctors' chamber. Hoffmann proposed that these courts of honour (of first instance) comprised five chamber members. Applications for initiating disciplinary procedures against a doctor were supposed to come from state authorities for medical and health matters, the public prosecutor or jointly from three doctors. It was proposed that the court of second instance or appeal court (*Ehrengerichtshof*) was made up of three members of the highest health authorities of the Reich and three members of the relevant doctors' chamber.[22]

In his comments on the draft ordinance, Hoffmann admitted that, personally, he had problems with the proposal of medical courts of honour and that he had to overcome his own reluctance to suggest such disciplinary institutions. However, he argued that the inclusion of provisions for professional discipline was necessary for a Doctors' Ordinance to stand any chance of receiving the support of the legislative powers of the Reich.[23] This justification was quite different from Graf's, who had portrayed the courts of honour as a desideratum of the medical profession itself rather than a requirement of the state.

In fact, the question of the need for medical courts of honour quickly became the bone of contention in discussions about a Doctors' Ordinance. An early opponent of Hoffmann's proposal was the physician and director of a hydropathic institute in Nassau, *Sanitätsrat* Karl Friedrich Ferdinand Runge (1835–1882). For him, the suggestion of establishing courts of honour was both unnecessary and dangerous. Integration of doctors into the voluntary medical societies was in his view sufficient to maintain standards of professional conduct. To ask for state-controlled disciplinary tribunals meant for him inviting reactionary political powers into the medical profession, which would suppress the freedom of doctors. French, Belgian, Dutch and English doctors, he predicted, would ridicule their German colleagues for asking 'for an extra stick from the government'. Moreover, Runge was concerned that unconventional innovators might be disciplined by the courts of honour, as might be inexperienced young doctors who were initially unfamiliar with professional etiquette. In contrast, it was hardly conceivable that these courts would dare to discipline a *Geheimrat* or a Professor.[24] There were, however, also more moderate critics. For example, the physician Xaver Mestrum, a member of the Wiesbaden medical society, argued for 'councils' of honour (*Ehrenräte*) instead of courts of honour. As elected and state-recognised representative bodies for all doctors of a district, the councils would intervene and mediate in cases of intra-professional conflict. Only in cases where this was unsuccessful would they reconstitute themselves as courts of honour with disciplinary powers.[25] Many medical societies actually already had councils for arbitration, so Mestrum's suggestion went only a small step further.

In July 1881, at the ninth annual meeting of the German Association of Doctors' Societies in Kassel, the issue of courts of honour was presented by the district physician of Hamburg-Altona, Julius Wallichs (1829–1916). He cited a survey of 92 German medical societies, with a total membership of approximately 6,000 doctors. Fifty of these societies had some form of disciplinary or arbitration panel, and a total of 32 cases had been reported by them. Wallichs suggested that this relatively small number of cases reflected a desirable effect of the recent campaign for collegiality in the medical societies and of the mere existence of such panels.[26] The tenth *Ärztetag*, held in Nuremberg in 1882, generally approved of Hoffmann's principles for a Doctor's Ordinance with a clear majority of 69 against 8 votes. However, an important amendment was made to the section on medical courts of honour. While Hoffmann had envisaged that these were to be formed by state-authorised doctors' chambers, the assembly merely agreed that such courts could be formed, voluntarily and from their own membership, by the various medical societies. These would be the courts of first instance, with an entitlement to decide on the exclusion of a member from the society. On the composition of courts of second instance (appeal courts) no agreement was reached at all, so that the form of these latter courts was left to the discretion of the medical societies, even though it was envisaged that these higher courts would also deal with cases involving doctors who were not members of a medical society.[27] The implication of this vote of the *Ärztetag* was that the majority of the organised medical societies preferred a system of professional self-regulation without state control, which therefore could only cover the voluntary members. This meant that members accused of professional misconduct could simply avoid disciplinary procedures by leaving their society. The envisaged courts of second instance, also covering non-members, had in fact no legitimisation and were therefore hardly more than wishful thinking at this stage.

The debate on a Doctors' Ordinance and medical courts of honour also entered the lower house of the Prussian Parliament (*Preußisches Abgeordnetenhaus*) and the German *Reichstag*. In a session of the Prussian Parliament in 1882, Graf, who was a member of the National Liberal Party, asked the Minister for Religious, Educational and Medical Affairs to promote the organisation of the medical profession and to submit a draft of a Doctors' Ordinance to the *Reichstag*. This motivated the eminent Berlin professor of pathology, Rudolf Virchow (1821–1902), who was an MP for the left-liberal German Progress Party, to speak out against this plan. Virchow challenged the view that the majority of the medical profession really wanted such tight organisation. Neither strict regulation of medical practice as a trade nor a civil-servant-like status of doctors would find sufficient support.[28] However, discussing a reform of the Trade Ordinance in 1883, the *Reichstag* showed understanding for a petition of the German Association of Doctors' Societies and adopted a resolution to ask Reich Chancellor Bismarck to submit a draft law on a Doctors' Ordinance, including provisions for medical courts of honour.[29] In the following three years, Graf used the annual debate on the health budget in the Prussian Parliament to remind the Minister, Gustav von Goßler (1838–1902), of the medical profession's request for state-authorised doctors' chambers. Even Virchow agreed eventually that the profession required some kind of organisation through the state. Following a direct petition by Graf to Bismarck as President of the Prussian State Ministry, a Royal decree on the establishment of medical representative bodies in Prussia was finally issued on 25 May 1887. It was accompanied by a ministerial issue in which von Goßler justified the measure by alluding to good experiences in other German states (namely Saxony, Bavaria, Wurttemberg and Baden) and its usefulness for organising medical advice in matters of public health.[30]

The Royal decree of 1887 constituted an important first step in the state organisation of the Prussian medical profession. It resulted in the creation of a doctors' chamber in each of the 12 Prussian provinces. One representative and one deputy for every fifty medical men were elected into a chamber. Its jurisdiction covered every doctor who was resident in the relevant province. While the chambers were chiefly representative bodies for professional interests and advisory boards for public health policies of the state, they were also granted some limited disciplinary powers. Doctors who had 'violated the duties of their profession considerably or repeatedly' or who, through their behaviour, had 'proven themselves unworthy of the respect that their profession demanded', could be punished with temporary or permanent withdrawal of their active and passive voting rights for the chamber. The accused had to be heard and a representative of the provincial government had to be involved as an advisor before the decision about a disciplinary punishment was arrived at. Appeals could be made directly to the Minister for Medical Affairs. These regulations did not apply to medical civil servants and medical military officers, who were already subject to their own disciplinary courts.[31]

Such limited disciplinary powers of the doctors' chambers were, however, apparently insufficient in the face of an expanding medical profession[32] and fierce competition among practitioners for contracts with the sickness insurance funds, leading to underbidding, excessive advertising, denigration of colleagues and poaching of colleagues' patients. Following the request of the *Reichstag*, Bismarck had asked, in December 1883, the governments of the various German states for views on a state organisation of the medical profession, but then had taken no further action. On behalf of the executive committee of the German Association of Doctors' Societies, Graf therefore submitted a petition to Bismarck in March 1889, asking directly for a Doctors' Ordinance for the German Reich.[33] The Chancellor's reply, on 3 May of the same year, was a bitter disappointment for Graf and other campaigners for professional discipline. Bismarck stated that it was a matter for the individual German states to extend the powers of their medical representative bodies (i.e., of the chambers), if that was deemed necessary; but he saw no need for the Reich to grant a uniform legal organisation to the medical profession in order to provide a basis for courts of honour.[34] This position of the Chancellor meant a major setback for the plans of the German Association to introduce a national Doctor's Ordinance including state-authorised medical courts of honour throughout the Reich.[35]

The strategy of Graf and the German Association during the years following Bismarck's rejection of a Doctor's Ordinance appears to have been threefold: promotion of codes of professional conduct (*Standesordnungen*) for the medical societies; support for the voluntary creation of courts of honours in the societies and campaigning for enhanced disciplinary powers of the doctors' chambers. In June 1889, after lengthy discussions, the seventeenth *Ärztetag*, meeting in Brunswick, adopted a number of 'Principles of a Medical Professional Code'. They were intended to serve as a model for codes of the member societies of the Association, which at this time comprised 219 societies with a total of 10,557 doctors. The Brunswick Principles prohibited medical advertising, including misuse of specialist titles, announcement of free treatment (particularly with a view to policlinics), underbidding in making contracts with sickness insurance funds, offering payments or benefits to third persons (e.g., midwives, nurses, pharmacists) to increase one's medical practice, prescribing or recommending 'secret remedies' (i.e., proprietary medicines of undisclosed composition), any attempt to attract new patients from a colleague's practice (especially during locum practice or as a consultant) and public disparagement of colleagues.[36] These rules had partly been taken from the existing professional codes of the medical district societies of Munich (1875) and Karlsruhe (1876).[37] The Munich society's code was in turn a translation of the 1847 'Code of Ethics' of the American Medical Association.[38] All of the Brunswick Principles directly or indirectly served the purpose of defusing competition within the expanding medical profession, educating incoming young practitioners about professional etiquette and demarcating the conduct of doctors from that of lay healers. To enforce these principles, the *Ärztetag* demanded that the medical societies created courts of honour or similar institutions, which could issue warnings or, as the ultimate sanction, decide to break-off all professional relations with the offender.[39]

In fact, a compilation published by Graf in 1890 listed 352 medical societies in the German Reich, of which 135 had a professional code (*Standesordnung*) and 194 some sort of disciplinary tribunal (*Ehrenrat*).[40] It is doubtful however whether the social sanctions of these voluntary societies had any major effect on those doctors who regarded medical practice as a free, competitive business or who dropped any considerations for their colleagues when they had to face economic problems themselves.[41] The disciplinary powers that had been granted to the Prussian medical chambers with the Royal decree of 1887 were applied, but only in a small number of cases, and there were indications that withdrawal of the right to vote (and to be elected) for the chamber was an inadequate punishment. By the end of 1891, only 52 cases had been dealt with under these regulations in the whole of Prussia. Of these, 10 had resulted in withdrawal of the active and passive voting right and 12 in a warning; 8 cases were not closed yet, and in 21 cases withdrawing the right to vote had been described as inappropriate – this sanction being regarded as either too mild or as too severe.[42]

In June 1890, the eighteenth *Ärztetag*, held in Munich, passed a resolution to increase the disciplinary powers of the doctors' chambers and medical district societies, especially in view of professional misconduct linked to contract practice under the national sickness insurance scheme. In October of the same year, the newly founded executive committee of all Prussian medical chambers arrived at the same conclusion, and several Prussian medical societies directly petitioned the Minister for Religious, Educational and Medical Affairs to the same effect.[43] Von Goßler's successor as Minister, Count Robert von Zedlitz-Trützschler (1837–1914), eventually responded to these medical requests. In January 1892, he issued an instruction to all Prussian provincial governments to invite the doctors' chambers to submit their expert opinions on the proposal to create medical courts of honour like those for lawyers.[44] This ministerial instruction initiated the second important step, after the Royal decree of 1887, towards a state organisation of medical discipline. All 12 doctors' chambers in Prussia agreed in principle with this proposal, which had basically brought Hoffmann's suggestions of 1880 back on the agenda.[45]

However, a critical minority within the profession, chiefly represented by the Medical Society of Frankfurt/Main and by the medical members of the Prussian Parliament (*Abgeordnetenhaus*) Rudolf Virchow and Paul Langerhans (1820–1909), again voiced objections. First, it was argued that the whole idea of applying the disciplinary regulations for lawyers to the medical profession was flawed. Lawyers could be regarded as mediate civil servants, because legal representation in the courts was compulsory. The vast majority of doctors had no comparable functions in relation to the state – except the medical civil servants and medical military officers, who were going to be exempted anyway from the proposed legislation. Without specific state duties for doctors, a disciplinary jurisdiction that would be authorised and controlled by the state seemed unjustified. Second, there were fears that patients and their relatives might overwhelm the courts of honour with unjustified complaints and thus damage the reputation of medical practitioners. Finally, there were concerns that these courts might become instruments of political control by the state. These latter worries arose in particular from the fact that the Lawyers' Ordinance had placed conduct within as well as *outside* professional practice under disciplinary control.[46]

The doctors' chambers wanted those medical men who served as civil servants or military officers to be included under the new disciplinary regulations. However, von Zedlitz-Trützschler's successor as Minister for Religious, Educational and Medical Affairs, Robert

Bosse (1832–1901), refused and threatened to drop the whole plan of courts of honour for doctors if the chambers continued to insist on this matter.[47] It was clearly unacceptable to the state that officers and civil servants, two professional groups at the top of the hierarchy of Wilhelmian society, should be judged by a court consisting mainly of medical practitioners. Further concerns in the ministry were raised by a letter sent by Eduard Graf as chairman of the medical chamber for the Rhine Province in November 1894. Waiting for a draft bill on courts of honour from the ministry, Graf – himself, as mentioned above, a member of the National Liberal Party, and in 1894 vice-president of the Prussian Parliament – had become impatient. He had also been incensed by an advertisement of a social-democratic doctor, who promised to colleagues 'with political affinity to the working class' lucrative contracts with sickness insurance funds.[48] For representatives of the medical profession such as Graf the links between the boards of the insurance funds and the Social Democratic Party had increasingly become a concern. In his letter Graf warned the Minister that 'the poison of Social Democracy' was infiltrating the medical profession. Dependent on contracts under the state's compulsory health insurance scheme for workers, doctors were beginning to lean, Graf suggested, to where the power lay. A 'tighter organisation' of the profession was therefore necessary. The letter was co-signed by the chairmen of the doctors' chambers for Brandenburg–Berlin, Julius Becher (1842–1907), and for Schleswig-Holstein, Julius Wallichs (1829–1916). Published in the *Ärztliches Vereinsblatt für Deutschland* in the following year,[49] it quickly evoked fears that doctors sympathetic to Social Democracy were going to be disciplined.[50] When the letter's contents became a topic in the daily press, Bosse took action and ordered his ministry to draft a bill on medical courts of honour within six weeks.[51] The draft was eventually made public in early 1896.[52]

It still took almost three years of negotiations between the medical chambers and the ministry before the bill was presented to the Prussian Parliament.[53] The problem concerning doctors who were civil servants or military officers was solved by excluding them also from membership of the planned courts of honour. In this way they could neither be judged by these courts nor sit on them as judges, but were subject to their own disciplinary tribunals. However, fears of political control continued. Against initial resistance from the ministry, the medical chambers eventually succeeded in putting through a clause stating that a doctor's 'political, scientific and religious views and actions as such' could never be subjected to disciplinary proceedings.[54] Besides worries that social-democratic doctors might be disciplined, there had also been concerns that the courts of honour might target doctors who practised homoeopathy or naturopathy.[55] The point on 'religious views' was probably included to prevent anti-Semitic sentiments from becoming effective, although there seems to have been no open discussion about this.

In 1899, both houses (*Abgeordnetenhaus* and *Herrenhaus*) of the Prussian Parliament accepted the bill, which was signed by Wilhelm II on 25 November of that year and became law on 1 April 1900.[56] In each of the twelve Prussian medical chambers a court of honour was formed, which consisted of four of the chamber's doctors and a professional judge from a general court. The latter acted as investigating officer, and the role of prosecutor was taken by a representative of the provincial government. The crucial section 3 of the courts of honour law demanded two things of a doctor: that he practised his profession conscientiously; and that he showed himself worthy of the respect that his profession demanded, in his conduct both during and outside medical practice. This was virtually the same wording as in the equivalent section 28 of the Lawyers' Ordinance.[57] Although several medical chambers drew up more specific professional codes,[58] these were not legally binding.[59] The new courts of honour could issue warnings and reprimands, punish with fines of up to 3,000 Mark,[60] or withdraw (temporarily or permanently) the right to vote for or be elected to the chamber. Combinations between the various disciplinary measures were possible. Moreover, the courts of honour could decide to publish the case as a means to increase punishment. They were not entitled, however, to withdraw a doctor's practising license, the *Approbation*. This could only be done by the local state authorities, typically in cases of criminal conviction linked with temporary or permanent loss of civil rights.[61] Appeals could be made to a central Prussian Court of Honour (*Ehrengerichtshof*) for Doctors, located in Berlin. Its first chairman was the legally trained director of the medical department in the Ministry for Religious, Educational and Medical Affairs, Adolph Förster (1847–1919), who served in this function until his retirement in 1911.[62] Its members were four representatives from the medical chambers and two doctors directly appointed by the King.

The Prussian law on medical courts of honour thus reflected – with some relevant modifications – the proposals made by Graf and Hoffmann some 20 years earlier, though Graf did not live to see his success. He died on 19 August 1895.[63] The courts of honour legislation had been the product of different, but converging interests of the Prussian state and the medical profession. The state was interested in politically loyal and controllable doctors who guaranteed health care under the sickness insurance legislation and provided advice and support in matters of public health. Although Minister Bosse portrayed the bill during its parliamentary debate as resulting solely from the wishes of the medical profession,[64] internal evidence from the ministerial files points to the former interpretation. In the draft of his report to the King, dated 22 July 1899, Bosse emphasised that he was especially pleased that the two houses of the Prussian Parliament had accepted a formulation of section 3 in analogy to section 28 of the Lawyers' Ordinance, because in this way the behaviour of doctors *outside* medical practice also came under disciplinary control. A similar view was expressed in an internal report by the ministerial official, Altmann, who gave the hypothetical example of a social-democratic doctor who had made offensive remarks about 'highest persons' in an election meeting. In such a case, Altmann explained to the minister, both the criminal judge and the medical court of honour would have to take action.[65]

Leading representatives of the medical profession such as Graf, Becher and Wallichs, however, had sought backing by the state in order to deal with the problems of intra-professional competition and to strengthen the position and status of medical practitioners, especially in relation to the increasingly powerful sickness insurance boards. The new, state-authorised disciplinary tribunals, in which doctors, jurists and government officials cooperated, seemed thus to satisfy needs on both sides.

As mentioned above, Prussia had not been the first German state to create official disciplinary tribunals. Apart from the early forerunners Brunswick (1865) and Baden (1864/1883), also Oldenburg (1891), Hamburg (1894) and the Kingdom of Saxony (1896) had established structures for the disciplinary control of doctors. Unlike their Prussian colleagues, the Saxon doctors were given a legally binding professional code (*Standesordnung*) in addition to courts of honour, and in 1904 both these components were united in a Doctors'

Ordinance (*Ärzteordnung*) for the Kingdom of Saxony.[66] However, the campaigning of the German Association of Doctors' Societies and of the various medical chambers for state-authorised courts of honour was not equally successful elsewhere in the German Reich. Bavaria can serve as an important example of a German state in which efforts to achieve a state regulation for medical courts of honour faced, after initial progress, considerable obstacles and had in the end only rather limited success. These interesting Bavarian developments will be discussed in the following section.

1.3 The Bavarian medical chambers and their campaign for disciplinary powers

Medical chambers (*Ärztekammern*) as regional representative bodies for doctors who practised in Bavaria had been introduced 16 years earlier than their Prussian counterparts. A decree by the King of Bavaria in 1871 had established eight medical chambers, one for each of the Bavarian administrative districts. Composed of elected representatives from the various, voluntarily formed medical district societies (*ärztliche Bezirksvereine*), they met annually on the same day at the seat of their respective district government to discuss issues of professional organisation, medical practice and science and public health.[67] The chambers also sent their delegates to the enlarged Higher Medical Commission (*verstärkter Ober-Medizinalausschuß*), an advisory committee on matters of health policy to the Bavarian state government, that met annually in the Ministry of the Interior in Munich. Initially, the medical district societies had only been entitled to expel members that were seen as undesirable, but not to refuse any membership applications by local doctors. Following a proposal of the Bavarian medical chambers and its approval by the Higher Medical Commission, a new Royal decree, issued on 9 July 1895, enhanced the powers of the district societies. Membership in a medical district society remained voluntary and members could leave the society at any time. However, the societies were empowered to refuse admission and to discontinue membership under specific circumstances: if a doctor had lost his civil rights (i.e., after a criminal conviction); if he went bankrupt or if his conduct was judged as 'unworthy of the medical profession' and gave reason to expect that fruitful cooperation within the society would be impossible. Appeals against such a decision for exclusion from the district society could be made to the relevant doctors' chamber.[68]

As with the Prussian medical chambers from 1887, this was the start of formalised, though limited, disciplinary control of doctors, which targeted overly competitive behaviour among medical practitioners. Following the Royal decree of July 1895, leading representatives of the profession developed proposals for uniform standing orders for the medical chambers, which were then circulated among the eight Bavarian chambers in autumn of the same year.[69] Moreover, they suggested specific regulations that the chambers should recommend to the medical district societies for inclusion into their statutes.[70] The medical chambers should form special commissions, whose task was to deal with complaints (appeals) resulting from the regulations on exclusion from the district societies. This was in addition to existing chamber commissions who advised the state authorities in cases of potential withdrawal of the medical practising license. The district societies were supposed to form courts of honour or arbitration panels, which were to adjudicate in arguments between members and be invested with disciplinary powers. The courts of honour would be entitled to issue confidential warnings, confidential reprimands, public reprimands in the society's assembly, or, finally, to apply for exclusion of a member from the society. Members who had received a warning or reprimand would be able to complain to the society's assembly, which would then have to hear the honour court's view and make a decision on the case.

Those suggestions seem to have been well received in the medical chambers. The chamber for Lower Franconia and Aschaffenburg, for example, adopted them unanimously at its meeting in October 1895 and recorded in the minutes its explicit thanks to the proposals' main authors, Dr Friedrich Ernst Aub (1837–1900) of Munich and Dr August Georg Brauser (1833–1901) of Regensburg.[71] By 1896, uniform standing orders for the medical chambers, based on the proposals of the previous year, were in force.[72] In the chamber meetings of autumn 1896 a more extensive proposal for a Doctors' Ordinance (*Ärzteordnung*), following the example of the Kingdom of Saxony, was already being discussed. The idea was to develop a professional code (*Standesordnung*) that would be binding for *all* doctors in Bavaria, not just for those who were members of a medical district society. Such a step seemed necessary to the chambers, because they had no legal means of intervening in disputes with doctors who were *not* members of a district society.[73] In 1895, 81 per cent of the doctors of Bavaria were members of a medical district society, that is approximately 1 in 5 doctors practised outside the reach of the disciplinary mechanisms described above.[74] Making membership of a district society compulsory was suggested as one obvious solution to this problem, but a state-authorised Doctors' Ordinance, combining a binding professional code with structures of disciplinary control by the medical representative bodies was regarded as an alternative option, and a petition with both suggestions was subsequently sent by the chambers to the Bavarian government.[75]

The government invited a formal proposal, and in 1897 the chairmen of the eight medical chambers drew up a detailed 'Draft of a Professional Code for the Doctors of Bavaria'. In addition, they agreed on a general outline for regulations on medical courts of honour, including disciplinary procedures – a measure that was seen as necessary in order to enforce the professional code in practice.[76] Following their discussion at the chamber meetings in the autumn of 1897 and the following year,[77] the draft code and the outline on courts of honour were submitted to the Bavarian government. Outside of the medical chambers, though, severe criticism of the plans for a binding professional code was expressed. On 12/13 December 1897 the newspaper *Münchener Post und Augsburger Volkszeitung* ran a front-page article which fiercely attacked the politics of National Liberal doctors' leaders such as Graf, Aub and Brauser. Instead of enforcing a professional code, which would lead to frequent denouncements and to political and religious control, the medical representatives should rather fight for the admission of all doctors to practise under the sickness insurance system, for more doctors in rural and poor regions and for more school and ship doctors.[78] Still, the efforts of the Bavarian medical chambers to achieve a tighter, state-authorised professional control of medical practice, in order to curb competitive behaviour and misconduct, seemed to be heading towards success. Developments appeared to follow similar lines as those in Saxony and Prussia. In two sessions of the Bavarian Parliament (*Landtag*), on 25 and 26 February 1898, the preparations for official medical courts of honour with disciplinary powers were welcomed by several speakers.[79] One of them, MP Dr Heim, motivated the need for such courts with the 'dirty competition' which he believed to be prevalent among doctors and which he illustrated with drastic examples:

> With the sickness insurances it happens that doctors offer visits for 30 or 40 Pfennig, just in order to edge out a colleague. That is worse pay than what a luggage carrier [at the station] gets. It is clear then why, as it has happened in Munich…, a panel doctor delegated his visits of patients to a young medic, who hadn't even passed his *Physikum* [i.e., the preclinical exams] yet and hadn't taken yet any clinical courses, and didn't know anything about prescribing etc…[80]

Another of Heim's examples was that of a Prussian doctor who had bought a practice in a rural part of Bavaria, grossly overcharged the mostly poor inhabitants and then moved on to a new practice in another German state.[81] The planned disciplinary powers for the medical representative bodies, as other parliamentarians agreed, were expected to deal with such cases, and the Bavarian Minister of the Interior, Baron Maximilian von Feilitzsch (1834–1913), promised to take this matter further.[82]

In fact, the Bavarian state government forwarded the proposals of the medical chambers to the Higher Medical Commission, which approved them with minor modifications on 19 December 1898. Item No. 1 of the draft code obliged doctors – akin to the regulations for lawyers – to exercise their profession conscientiously and to 'maintain the honour and reputation of the profession through their conduct within as well as outside professional practice'. Responding to the concerns about potential political and religious control of doctors, a clause saying that 'naturally the religious and political behaviour is excluded' had been inserted under this item as well. Other key obligations of the professional code were that a doctor had to 'stand on the fundaments of medicine as taught at our universities'; to abstain from using differing views for advertising purposes; to support efforts in public health; and to oppose quackery and the trade with secret remedies. Several items of the code concerned the medical fee, aiming mainly at preventing undercharging and underbidding. Of particular significance was item No. 36, which obliged doctors to make contracts with the sickness insurance funds and other insurance organisations only through special contract commissions of their medical district society. This clause was relevant because a contract commission would have greater bargaining power vis-à-vis the insurance organisations than an individual medical practitioner. Moreover, underbidding among doctors could be avoided in this way.

The outline for the organisation of medical courts of honour placed the courts of first instance (*Ehrenräte*) in the medical district societies and the appeal courts (*Ehrengerichtshöfe*) in the medical chambers. The spectrum of disciplinary measures of those courts were supposed to comprise: (1) confidential warnings (2) confidential reprimands (3) public reprimands in the district society's assembly (4) fines up to 2,000 Mark (5) withdrawal of the active and passive voting rights for all of the societies' elections up to five years (6) temporary or permanent exclusion from consultations (7) increase of the last three forms of punishment by publication of the details in the press and (8) application for initiating procedures to withdraw the medical practising license (*Approbation*).[83]

By July 1899, the Ministry of the Interior had prepared a draft for a 'law concerning the medical professional code and courts of honour' and consulted with the Ministry of Justice for amendments.[84] The official draft bill was published on 28 September 1899.[85] According to this draft, the professional code and the medical courts of honour would cover all doctors practising in Bavaria. The committees of the medical district societies would be entitled to warn doctors whose conduct was seen as deviating from the professional code, which was supposed to be issued by a decree from the Ministry of the Interior. Unlike the original proposal of the medical chambers, the draft bill gave the district societies no right to create formal courts of honour, because (as the official 'reasons' stated),

> to make the district societies themselves a place for court of honour proceedings, and to impose on them the stigma that is linked with these, cannot be feasible, given the serious meaning that a disciplinary punishment has for a doctor, in view of the small number of members in some district societies, and especially the voluntary character of the formation of the district societies.[86]

Instead, the courts of honour of first instance were to be located in the medical chambers in the form of councils of honour (*Ehrenräte*), consisting of four doctors and a government administration official.

The disciplinary powers of these councils were supposed to encompass warnings, reprimands, fines between 20 and 2,000 Mark, and temporary or permanent exclusion from the district society. This spectrum of punishments closely resembled that in Saxony and Prussia, and – as in these two states – there was a provision for increasing punishment by publishing the verdict. As in Prussia and Saxony, the Bavarian courts of honour would *not* be entitled to withdraw a doctor's practising license – this remained a prerogative of the state authorities. The accused would be able to appeal to a court of honour of second instance in Munich, the *Ehrengerichtshof*, composed of eight delegates from the medical chambers (one from each) and a government official appointed by the Ministry of the Interior. As the 'reasons' of the draft bill stated, these regulations were thought to respond to the wishes of the Bavarian medical profession and to remove a defect of the current disciplinary mechanisms, which allowed a doctor to escape the professional code and discipline of his district society by quitting his membership (or by not joining a district society in the first place). Although influenced by both the Prussian and the Saxon example, the Bavarian draft bill deviated deliberately in one respect from the Prussian model, which had done without a binding professional code. Bavaria followed instead the example of the Saxon law of 1896 in combining disciplinary bodies with a code of professional conduct. The task of the Saxon courts of honour, which were formed in the medical district societies out of at least three members, was to decide about allegations of breaches of the professional code (and to arbitrate in quarrels between doctors or a doctor and another person).[87]

In its session of 30 October 1899, the Bavarian *Landtag* referred the draft bill – due to its specialist subject matter, without any general debate – to a parliamentary commission for preliminary discussion and amendment. Dr Friedrich Ernst Aub, who was present as a member of parliament, agreed with this action.[88] In fact, he was elected chairman of this commission, so that things seemed well in hand from the perspective of the organised doctors in Bavaria. Aub, the government's medical district officer for Upper Bavaria, was an influential representative of the medical profession and had become Graf's successor as president of the German Association of Doctors' Societies in 1897. In a preliminary meeting of the commission on 14 December 1899, Aub reported on the draft bill, and it was agreed to consider the professional code at the same time, as some of the code's points might be directly transferred into the planned law.[89]

However, Aub died on 18 March 1900.[90] The matter lost impetus and only in October 1901, after petitions by the medical chambers to the government and to the parliament, was the draft bill fully discussed in the commission for the first time. The role of speaker on the draft bill was given to MP von Landmann, the legally trained mayor of Günzburg (near Ulm), who was also a homoeopathic lay practitioner. It seems that these events were a significant factor in the subsequent developments which caused much irritation in the medical chambers and resulted in the failure of their plans for a Bavarian law on medical courts of honour.

Von Landmann regarded a professional code as superfluous and as incompatible with the freedom of medical practice as guaranteed in the Trade Ordinance of the Reich. More specifically, he argued that matters such as advertising, buying or selling a practice, medical fees, contracting with sickness insurance organisations, the choice of remedies or dealings with patients of other doctors, could not be subject to disciplinary proceedings in courts of honour, since these matters were not regarded as insults to one's honour from the perspective of non-medical, educated lay people. Although other members of the commission and Minister von Feilitzsch continued to be supportive of the professional code and draft bill as they had been proposed, considerable changes were made by the parliamentary commission. In particular, the rule that contracts with sickness insurance organisations could only be made by special commissions of the medical district societies, not by individual doctors, was deleted, as were the detailed regulations on medical fees.[91] The medical chambers submitted a further petition to the *Landtag*, asking for the original proposals of the government to be upheld, but to no avail, despite informal discussions of medical representatives with members of the parliament and in the Ministry of the Interior. Angry and frustrated, in January 1902, the representatives of the Bavarian medical chambers asked Minister von Feilitzsch to withdraw the draft bill altogether rather than to submit it to the plenum of the *Landtag* in the 'mutilated' form in which the parliamentary commission had left it.[92] The Minister withdrew the draft bill, and it was not discussed in the plenum.[93]

The medical chambers realised that they had lost their case, at least for the time being. For several years, discussions about state-authorised courts of honour and a legally binding professional code fell silent, both within the Bavarian government and the medical representative bodies. The matter was briefly raised in the plenary session of the Bavarian *Reichsratskammer* on 10 May 1904 with suggestions for a more generic medical professional code and for more government administration officials in the courts of honour, but Minister von Feilitzsch predicted that these ideas probably would not find the approval of the medical profession, which would have to make a new submission to the government anyway, for procedural reasons.[94]

It was only in 1908 that *Hofrat* Dr Wilhelm Mayer of Fürth, the chairman of the medical chamber for Middle Franconia, put the issue on the agenda again. In an article in the *Münchener Medizinische Wochenschrift*, including a letter in the name of the executive committee of the medical chambers to the Ministry of the Interior, he called for reforms in the court of honour proceedings at the district societies because of their apparent deficiencies.[95] The immediate circumstances were dramatic: Dr Moritz Hutzler, a physician at the *Gisela-Kinderspital*, a private paediatric clinic in Munich, had committed suicide because the court of honour of his local medical district society had found him guilty of 'a serious violation of his professional duties of collegiality' and had moreover stated that he also 'violated the principles of good faith in a general, civic sense'. Although the chairman of the court of honour, Dr Kastl, who was simultaneously chair of the Munich medical district society, had later made a personal declaration of honour for Dr Hutzler, thus qualifying the second part of the verdict, the latter shot himself on Easter Sunday 1907. After his death, documents in his favour were found among Hutzler's papers. The affair led to severe public criticisms of the court of honour concerned, which in turn tried in vain to defend itself through a civil libel action.[96] In light of this scandal, Mayer's reform proposals aimed at improving the existing disciplinary procedures by permitting the courts of honour at the district societies to request ordinary courts to question witnesses under oath and by involving jurists as legal advisors. He also recommended the creation of more formal appeal courts at the medical chambers and asked the government for advice on how to subject *all* doctors in

Bavaria to a professional code and courts of honour system. The continuing necessity of such disciplinary control was incontrovertible for him:

> That courts of honour...are necessary for the medical profession requires no proof. In no academic profession – including lawyers – are personal collisions, the temptation to gain advantages through lesser or greater duplicity, as well as the reluctance to submit to firm norms, more frequent than among the doctors.[97]

A colleague in Mayer's medical chamber, Hans Doerfler, responded with the suggestion to campaign only for state-authorised courts of honour, without a specific professional code, because this Prussian model might have greater chances of acceptance by the *Landtag*.[98] In fact, the Minister of the Interior had replied to Mayer and his colleagues that the medical chambers could agree for themselves on a common professional code, but that a state ordinance on courts of honour along the proposed lines would require legislation.[99] Following discussions in the medical chambers,[100] a draft professional code was submitted to the state government at the start of 1910. However, the Higher Medical Commission, which subsequently had to scrutinise the draft, insisted on deleting the crucial rule that contracts with the sickness insurance organisations should only be made through contract commissions of the medical district societies, and it also deleted a section that declared it as desirable for every doctor to be a member of a professional and economic organisation. Only in this revised and thus considerably weakened form did the Minister approve the code in April 1910, as a provisional basis for the decision-making in courts of honour proceedings.[101]

Similarly, the additional attempt to achieve an ordinance on courts of honour met resistance. In response to a draft submitted by the medical chamber of Upper Bavaria, which had the support of six of the eight Bavarian chambers, the Ministry of the Interior replied in June 1912 that it was too early for such an ordinance and that it was advisable to wait to see how the professional code (of 1910) worked in practice.[102] It soon seems to have become evident, however, that the old problems had not been solved. Responding to a petition from the medical chambers in late 1913, the Ministry of the

Interior prepared a detailed draft bill including the organisation of the medical district societies and chambers, a professional code and regulations on medical courts of honour. According to the 'reasons'-section of this draft bill, the heightened competition between doctors, caused by their increasing concentration in the cities and by the selective contracting of the sickness insurance organisations, made these disciplinary mechanisms necessary. Moreover, the examples of Prussia, Saxony, Baden and Brunswick, which, by then, all had medical representative bodies with legally authorised disciplinary powers, was quoted.[103]

However, the outbreak of the First World War seems to have prevented any further development of this draft bill. In the early years of the Weimar Republic, the old Royal decree of July 1895 continued to be the basis for the operation of the district societies and medical chambers.[104] It was only in 1927 that Bavaria legally introduced for its doctors compulsory membership in the local medical district societies, obligatory representation in a central, state medical chamber (*Landesärztekammer*) and professional courts for doctors – almost three decades after Prussia. A medical professional court (*ärztliches Berufsgericht*), consisting of four doctors and one legally trained member, was formed in each governmental district, and an appeal court (*Landesberufungsgericht*) consisting of five doctors and two judges was established in Munich. The disciplinary sanctions included warnings, reprimands, exclusion from the medical district society, fines up to 10,000 Mark and in special cases publication of the details.[105]

This 'story' of the Bavarian medical profession's campaign for state-authorised disciplinary structures was thus – for the period of the *Kaiserreich* – one of disappointments and only very limited successes. Representatives of the profession tended to portray the reasons for this in a personalised manner, blaming in particular the failure of the draft bill of 1899 on MP von Landmann, who, as a jurist and lay practitioner, was accused of a 'lack of understanding of medical ethics' and 'disrespect for the science of medicine'.[106] However, ultimately the medical representative bodies had not enough political influence in Bavaria, it seems, and the Ministry of the Interior had only been half-heartedly supportive. The transition to a state-authorised disciplinary jurisdiction occurred later than in most German states during the Weimar Republic.

1.4 The activities of medical courts of honour

In Prussia, in contrast, the new, state-authorised courts of honour quickly displayed a lively activity in disciplining doctors. Thanks to surviving detailed records of the Prussian ministerial bureaucracy, it is possible to characterise these activities on a quantitative basis. In addition, the Prussian Court of Honour for Doctors, that is the appeal court in Berlin, regularly published collections of its decisions that were thought to be of substantive or procedural importance.[107] Such decisions were likewise published in a ministerial periodical, the *Ministerial-Blatt für Medizinal- und medizinische Unterrichts-Angelegenheiten*, and in medical professional journals, such as the *Ärztliches Vereinsblatt für Deutschland*, the *Berliner Ärzte-Correspondenz* and the *Münchener Medizinische Wochenschrift*. On this basis I will discuss which issues of professional conduct kept the medical courts of honour busy, how they explained their decisions and to what extent the fears of the critics turned out to be justified.

The 12 provincial medical courts of honour (attached to the doctors' chambers) in Prussia had to provide annual reports of their activity, which were collected in the Prussian Ministry for Religious, Educational and Medical Affairs and, from 1911, the Prussian Ministry of the Interior. These returns provide information about the annual number of allegations of medical misconduct received by the courts of honour, the sources of these allegations, the number of cases in which the doctor concerned was 'punished' (from warnings and reprimands to fines and withdrawal of the voting right) and the total number of doctors living in Prussia within the jurisdiction of the courts of honour system. These data are available from 1903, and I have compiled them for the remainder of the *Kaiserreich*, including the years of the First World War, and the early years of the Weimar Republic.[108]

As shown in Table 1.1, in the pre-war period the 12 Prussian courts of honour received approximately 500–600 allegations of medical misconduct per year. In relation to the number of doctors who were subject to their jurisdiction, this meant that approximately 3 or 4 allegations for every 100 doctors were made each year. During the war, the number of accusations of misconduct fell to approximately 2 per 100 practitioners – a decrease which can probably be explained

Table 1.1   Activity of the Prussian medical courts of honour, 1903–1921

Year	Allegations (A) received (A/D%)	Punishments (P) imposed (P/D%)	Doctors (D) within jurisdiction
1903	492 (3.3)	159 (1.1)	14,740
1904	568 (3.7)	151 (1.0)	15,332
1905	497 (3.1)	177 (1.1)	15,848
1906	525 (3.3)	122 (0.8)	15,816
1907	482 (3.0)	146 (0.9)	16,107
1908	551 (3.4)	167 (1.0)	16,222
1909	530 (3.2)	122 (0.7)	16,333
1910	554 (3.4)	130 (0.8)	16,305
1911	551 (?)	156 (?)	?*
1912	571 (3.4)	122 (0.7)	16,713
1913	607 (3.5)	152 (0.9)	17,173
1914	619 (3.5)	115 (0.7)	17,486
1915	305 (1.7)	51 (0.3)	17,526
1916	286 (1.8)	80 (0.5)	16,308
1917	300 (1.9)	72 (0.5)	15,587
1918	282 (1.8)	43 (0.3)	15,797
1919	370 (1.9)	48 (0.3)	18,986
1920	638 (3.3)	87 (0.5)	19,085
1921	685 (3.4)	131 (0.6)	20,257

Note: *no figure for Rhine Province for 1911.

Source: Data compiled from GStAPK I. HA Rep. 76 VIII B, Nr. 782–783.

by the fact that many doctors were then in active service and a proportion of disciplinary cases may have been dealt with by military tribunals instead. This explanation is also supported by the fact that the number of received allegations per 100 doctors rose again to the pre-war level of over 3 in the early 1920s. In the period before the First World War, approximately 1 in 3 accused practitioners were given a disciplinary sentence; during and after the war the proportion of punished cases halved. The latter decrease in punishments can be explained with the fact that a general amnesty for disciplinary offences was pronounced after the war, in 1919, so that many sentences issued during the war were suspended.[109]

If one assumes that there were only a few doctors who were accused of misconduct more than once, quite a considerable number of medical practitioners must have faced involvement in courts of honour proceedings by the start of the First World War. According to the published official sanitary reports for Prussia from 1903 to 1913, the 12 provincial courts of honour of first instance (i.e., those attached to the doctors' chambers) dealt with 4,415 matters in informal proceedings (which could result in warnings, reprimands and fines up to 300 Mark), 1,357 matters in formal proceedings (which could result in fines up to 3,000 Mark and withdrawal of the active and passive voting rights in the chamber elections) and 676 matters in mediation proceedings.[110] These figures should be seen in light of the total number of doctors who were resident in Prussia and subject to the medical chambers' disciplinary jurisdiction. In the period from 1903 to 1914, their number rose from 14,740 to 17,486 (see Table 1.1). In the 12 years from 1903 to 1914, the Prussian medical courts of honour reported a total of 1,719 disciplinary punishments. If one ignores multiple offenders and changes in the composition of the profession through deaths and new entries, as many as 1 in every 10 Prussian medical practitioners may have been disciplined by a court of honour by the beginning of the First World War.

Doubtlessly then, the Prussian medical courts of honour were busy institutions. But who kept them so busy, and what types of (mis-) conduct gave reason for disciplinary punishment? The ministerial records provide a rough breakdown of the allegations of misconduct according to their source (see Table 1.2). The fears that many patients or their relatives would make complaints turned out to be justified: approximately 20 to 30 per cent of the allegations came from 'private persons'. Also the worries that the state wanted to exercise control over doctors' behaviour proved to be well founded: another 20 to 30 per cent of the accusations were submitted to the courts of honour by 'the authorities'. The majority of allegations of misconduct, however, were made by other doctors, reflecting a strong culture of denouncement and self-discipline within the profession. In the early 1900s, shortly after the courts of honour had started to operate, the proportion of this source of accusations was over 50 per cent, then levelling at approximately 40 per cent. It dropped during the war to approximately 20 to 30 per cent, only to rise again to pre-war levels in the early years of the Weimar Republic. Before the war, approximately 5 per cent of the cases were initiated by the accused practitioners themselves in order to clarify an issue or to clear their names – an indication that the courts of honour were widely

Table 1.2 Sources of allegations of misconduct received by the Prussian medical courts of honour, 1903–1921

Year	Private persons (%)	The authorities (%)	Other doctors (%)	The accused (%)	Anonymous (%)	n =
1903	15.2	20.3	58.1	5.3	1.0	492
1904	21.8	18.7	52.8	4.6	2.1	568
1905	21.9	24.9	45.9	6.4	0.8	497
1906	22.9	25.7	46.7	3.6	1.1	525
1907	24.5	27.4	39.8	6.6	1.7	482
1908	20.7	28.7	44.3	5.4	0.9	551
1909	22.3	25.5	46.2	5.1	0.9	530
1910	33.9	26.7	33.4	5.4	0.5	554
1911	29.2	27.0	36.7	5.8	1.3	551
1912	28.5	24.2	41.3	4.9	1.1	571
1913	25.5	23.2	45.0	5.3	1.0	607
1914	26.0	31.7	36.0	5.3	1.0	619
1915	26.2	46.2	23.3	3.3	1.0	305
1916	29.7	42.0	23.1	4.9	0.3	286
1917	26.3	37.0	32.7	3.7	0.3	300
1918	31.2	35.1	29.8	2.1	1.8	282
1919	27.8	28.9	38.6	3.5	1.1	370
1920	25.2	27.4	42.3	3.8	1.3	638
1921	19.1	29.3	46.4	4.2	0.9	685

Source: Compiled from GStAPK I. HA Rep. 76 VIII B, Nr. 782–783.

recognised by the profession as the authoritative arbiters in questions of correct conduct. Approximately only 1 per cent of accusations were received from anonymous sources.

The large proportion of accusations made by other doctors already indicates that the problem of fierce competition between practitioners, the key issue which had been highlighted in the debates about the need for courts of honour, played a main role. The Ministry for Medical Affairs did not require of the courts of honour to provide figures on the distribution of cases over particular types of medical misconduct. Only the types of disciplinary offences as such, which a court of honour had punished in the course of the year, had to be mentioned in its annual report to its provincial government. The medical court of honour for Brandenburg and Berlin performed this task very diligently, so that the surviving reports (which were forwarded to the Ministry) for the years 1903 to 1920 collectively give a picture of the most frequently named disciplinary offences. In terms of the development of the numbers of allegations received and punishments imposed, the Brandenburg–Berlin court of honour broadly followed the collective trends of the 12 Prussian courts of honour (see Table 1.3). For the period from 1903 to 1914 one finds an average of 4.5 allegations of misconduct per year per 100 practitioners living in Berlin or Brandenburg under the court's jurisdiction, a figure which is somewhat higher than the annual average of 3.3 complaints per 100 doctors for Prussia as a whole. However, the ratio of disciplinary punishments to the number of doctors was about the same in that period, that is approximately 1 per 100 practitioners (0.8 for Brandenburg–Berlin, 0.9 for Prussia). The higher number of allegations for Brandenburg–Berlin can probably be explained by the greater concentration of doctors practising in the capital and thus greater competitiveness among them.

Table 1.3 Activity of the medical court of honour for Brandenburg and Berlin, 1903–1920

Year	Allegations (A) received (A/D%)	Punishments (P) imposed (P/D%)	Doctors (D) within jurisdiction
1903	158 (4.4)	35 (1.0)	3,591
1904	166 (4.6)	19 (0.5)	3,607
1905	163 (4.3)	31 (0.8)	3,819
1906	180 (4.5)	31 (0.8)	3,992
1907	166 (4.4)	45 (1.2)	3,807
1908	182 (4.8)	46 (1.2)	3,812
1909	198 (5.1)	30 (0.8)	3,908
1910	188 (4.8)	28 (0.7)	3,925
1911	161 (4.0)	28 (0.7)	3,968
1912	181 (4.4)	21 (0.5)	4,125
1913	183 (4.4)	24 (0.6)	4,159
1914	179 (4.2)	23 (0.5)	4,222
1915	110 (2.6)	14 (0.3)	4,197
1916	102 (2.4)	21 (0.5)	4,199
1917	108 (2.6)	25 (0.6)	4,143
1918	103 (2.5)	22 (0.5)	4,196
1919	135 (3.0)	12 (0.3)	4,469
1920	169 (3.2)	26 (0.5)	5,225

Source: Compiled from GStAPK I. HA Rep. 76 VIII B, Nr. 830.

As can be seen from Table 1.4, a relatively small number of types of disciplinary offences dominated in the reports of the Brandenburg–Berlin court of honour. Clearly, problems in the professional relationships between doctors, rather than in the doctor–patient relationship, led most frequently to disciplinary punishments. Most prominent by far was the issue of excessive advertising, which was regarded as dishonourable and unworthy of a doctor, since it was common practice amongst 'quacks'. The fact that many doctors nonetheless 'offended' in this area can be seen as a result of strong competition among medical practitioners. Some doctors also insisted that the freedom of medical practice guaranteed in the Trade Ordinance permitted them to advertise their services as they liked. However, already in May and July 1901, and again in December 1902, the Prussian Court of Honour for Doctors, that is the central appeal court in Berlin, had issued decisions which established the principle that blatant advertising constituted a violation of the professional honour of doctors and that disciplinary punishment for this offence did not contradict the Trade Ordinance.[111] Other frequent disciplinary offences, such as financial misconduct, slander or libel, practising from different places and the poaching of other doctors' patients, can likewise be interpreted as manifestations of fierce competition in the profession.

Refusal of medical assistance, however, was sometimes justified with the provisions of the Trade Ordinance, which in principle allowed this. However, in December 1902, the Prussian Court of Honour for Doctors ruled that such refusal of help was punishable in cases of acute danger to the patient's life.[112] This decision correlated to the requirement for any citizen to provide help in emergencies according to the German Penal Law (see above), but it also recognised a special professional 'duty to assist' for doctors. Against the spirit of the Trade Ordinance, the courts of honour tried to create sharp distinctions between medical doctors and lay practitioners. Any business links with non-licensed healers, especially naturopaths, were branded as 'dishonourable as such' and were punished, as was any co-operation with lay societies for natural healing methods.[113] Moreover, the difficulties of doctors in relation to the sickness or accident insurance funds[114] were reflected in the court of honour decisions. Not only did they give rise to 'unfair competition' and 'lack of collegiality' between doctors competing for patients under

Table 1.4 Reasons for disciplinary punishments cited by the medical court of honour for Brandenburg and Berlin, 1903–1920

Reason for punishment	Cited n-times
Advertising	72
Financial misconduct	30
Slander or libel	29
Sexual offence	23
Lack of collegiality	21
Links with non-licensed healers	18
False certifying	13
Practising from different places	12
Refusal of medical assistance	11
Unfair competition	10
Assault	9
Fraud	9
Negligent certifying	9
Adultery	7
Breach of contract	7
Failure to provide medical report	7
Trade with secret medicines	7
Breaking of word of honour	6
Dishonourable conduct	6
Offensive letter to the authorities	6
Poaching other doctors' patients	6
Treatment by letter	5
Maltreatment	4
Offending public decency	3
Sale/purchase of practice	3
Unauthorized use of title	3
Breach of professional secrecy	2
Ignoring official order/request	2
Dispensing to patients	2
Theft or embezzlement	2
Blackmail	1
Blank prescriptions	1
Culpable homicide	1
Obstructing an officer	1

Source: Compiled from GStAPK I. HA Rep. 76 VIII B, Nr. 830.

the insurance system, but they also involved negligent and false certifying, and failure to provide medical reports (cf. Table 1.4).[115] The medical courts of honour also acted as enforcers of professional solidarity among doctors versus the insurance boards. The medical chambers and the Association for the Doctors of Germany for the Protection of their Economic Interests (the so-called *Hartmann-Bund*, founded in 1900) made their members give their word of honour that they would not enter a contract with a health insurance organisation without the permission of their contract commissions. If that promise was broken, the medical courts of honour pursued disciplinary proceedings against the doctors concerned, especially in periods of dispute with the insurance boards.[116]

Within the doctor–patient relationship, sexual offences involving female patients were a major reason for disciplinary punishment. A doctor's sexual misconduct was also disciplined by the courts of honour when it had occurred outside medical practice.[117] Overall, however, disciplinary cases arising from disputes within the profession predominated. This can also be concluded from the decisions of the Prussian Court of Honour for Doctors (*Ehrengerichtshof*) which were regularly published by the Ministry. As shown in Table 1.5, the majority of these decisions concerned intra-professional issues, such as excessive advertising, relations to non-licensed healers, unfair competition in contracting with health insurance funds, insults against other doctors and lack of collegiality, disregard for agreed levels of fees or selling a practice for an exaggerated price.[118] Typically for a new institution, many decisions of this appeal court clarified procedural questions. In comparison

Table 1.5 Published decisions of the Prussian court of honour for doctors, 1900–1914

Area of disciplinary case	Number of decisions
Advertising	31
Insults against colleagues or lack of collegiality	25
Competition with other doctors	22
Contracting with health insurance funds	21
Financial misconduct	13
Links with non-licensed healers	11
Doctor–patient relationship, practice of medicine	25
Conduct outside medical practice	8
Procedural matters	83

Source: Compiled from Ministerial-Blatt für Medizinal- und medizinische Unterrichts-Angelegenheiten 1 (1901) – 14 (1914).

with this, the number of published decisions on matters regarding the doctor–patient relationship was small. Most of these dealt with cases of negligence, particularly in providing certificates, but there were also decisions concerning refusal of medical assistance in emergencies, sexual offences against patients, breaches of confidentiality[119] and malpractice.[120]

Decisions regarding doctors' conduct outside medical practice included, as foreseen by critics of the court of honour system, some political cases. For example, a social-democratic doctor was disciplined because he had accused the police of inhuman actions against workers.[121] Another doctor was punished for having sent an 'offensive postcard' to his provincial government, complaining about a lack of 'Germanness' in a local colleague of Polish descent.[122] In a number of decisions the Prussian Court of Honour for Doctors made clear that political, scientific and religious actions could in fact be subject to disciplinary punishment if their *form* was considered offensive, mean or dishonourable.[123] On this basis, doctors practising alternative healing methods were also disciplined if they had offended colleagues as part of their criticism of orthodox, scientific medicine or if they had established links with naturopathy societies that were known for their 'hostility towards doctors'.[124]

The widely published decisions of the Prussian Court of Honour for Doctors not only provided precedents for the decision-making of the disciplinary tribunals of first instance at the medical chambers but also formed – collectively – a kind of professional code of conduct. Over the years, these decisions thus fulfilled the role of a legally binding code of medical ethics. As a contemporary commentator put it, the jurisdiction of the Court of Honour for Doctors would gradually develop 'a medical law of honour' (*ein ärztliches Ehrenrecht*).[125] What kind of ethics was expressed in the judgments of courts of honour, and what considerations guided the judges of those courts in their decision-making? The practitioner-centred rather than patient-centred character of the disciplinary actions of such courts has been shown above. In the following section, I will further explore the underlying notions of medical professional ethics through closer examination of some specific cases and by drawing upon Georg Simmel's and Pierre Bourdieu's concepts of honour.

1.5 Professional ethics, honour and discipline

Particularly revealing in this respect are cases in which patients' interests competed with the professional interests of doctors. This can be illustrated, for example, by a disciplinary case from the Prussian province of Westphalia in 1905. A medical practitioner had been asked for help by a day labourer, who claimed that his daughter had been subjected to 'indecent manipulations' by the district physician during a physical examination. The practitioner assisted the labourer by writing a report for him and sending it to the public prosecutor. Preliminary proceedings against the district physician were subsequently initiated, but later abandoned because of lack of evidence. The medical practitioner, however, received a reprimand by the provincial medical court of honour for having committed an offence against the duty of collegiality, and he also had to bear the costs of the disciplinary proceedings. It was wrong, stated the court of honour, to have trusted the allegations of an uneducated man against a colleague and to have reported the colleague without giving one's own name.[126] Apart from the obvious class issue involved in this case, this outcome demonstrated that concern for the profession's public reputation prevailed over concern for patients' interests.

Similar attitudes were reflected in another case which occurred a few years later in Berlin. A doctor from Güstrow/Mecklenburg had been brought to the attention of the Berlin police by the Dutch Consulate General, because he regularly saw patients in a Berlin hotel, where he treated them with his special method of formic acid injections, for example as a therapy for cancer. The consulate asked the police whether it issued warnings against this therapeutic method, and if so, whether such warnings applied to the method itself or the manner in which this doctor practised it. As it turned out, the medical court of honour for Brandenburg and Berlin had reprimanded this doctor in 1908 – not because of his unusual therapeutic method but because of his practising from two different places, that is his home town and the hotel in Berlin, which was considered unworthy of a doctor. However, he had successfully appealed against this decision, as it could not be disproved that he only saw patients in the hotel with whom he had previously made appointments because Berlin was a more convenient place for them.

He had thus been acquitted, and the police had not issued a warning to the public. The Minister for Medical Affairs, who was consulted by the Berlin Chief of Police, merely advised the latter to tell the Dutch consulate that it was at liberty to approach him via the official diplomatic channels if it would like to know about the therapeutic method.[127] Again, the handling of this case thus reflected greater concern about professional reputation than about the safety of patients.

The concern for the profession's reputation in the eyes of the public, and for maintaining the honour of doctors, went so far in some decisions of the medical tribunals that they provoked critical discussion within the profession itself. For example, in its decisions of 13 October 1903, the Prussian Court of Honour for Doctors confirmed the disciplinary conviction of a doctor who had given a voluntary discount on fees to those patients who paid them immediately. The same doctor had then also sold his medical practice. According to the court's reasoning, the discounts were 'highly unworthy of a doctor', because they put medical services on the same level as 'business goods'. Moreover, the doctor's defence that he had not been aware that selling his practice was against professional norms, did not protect him. As it was the case in criminal law, the court of honour argued, not knowing that one's action is a punishable offence does not count as an excuse.[128] In an article in the *Ärztliches Vereinsblatt für Deutschland*, a medical colleague from Münster/Westphalia, Dr Mulert, defended the disciplined doctor, arguing that there was nothing dishonourable in giving a discount to prompt payers and that a clause on such discounts should even be printed on all invoices. He also questioned the court's stance on selling a medical practice, since nobody blamed a businessman or a farmer for reaping the profits when giving up their well-run enterprise. If the successor was correctly informed about the usual income from the practice, he had no reason to complain. The editor of the *Vereinsblatt*, however, *Sanitätsrat* Dr O. Heinze, who was simultaneously general secretary of the German Association of Doctors' Societies, added sharp comments to Mulert's article, justifying the court's decisions: printing a discount clause on invoices would mean 'putting the doctor on the same level as that species of "business people" who use such hints in order to attract buyers'; and selling something 'unsaleable' like a medical practice was indicative of an 'improper motive'.[129]

Clearly, the position of the Prussian Court of Honour in this case as well as Heinze's defence of its decisions reflected the widespread animosity in the organised medical profession against the implications of the Trade Ordinance (see above). However, as this case further illustrates, there was also a more general problem with reconciling the specific requirements of *medical* professional honour with notions of honour in other areas of daily life. In fact, in 1901 another article in the *Vereinsblatt* had discussed the topic of 'professional consciousness' (*Standesbewußtsein*) and 'professional honour' (*Standesehre*), raising the question why it was, for example, considered 'unworthy of the profession' (*standesunwürdig*) if a doctor advertised his services repeatedly, but not if a businessman did the same, or if a doctor offered his services at a lower cost than a colleague when he was able to provide them more cheaply. The conclusion was that medical professional ethics included some specific duties in addition to those of general morality, and these specific requirements were designed to protect the reputation of the medical profession in the public and thus to secure patients' trust in their doctor. These special 'doctor's ethics', argued the medical author, should therefore already be taught in the medical curriculum of the universities, and later be cultivated in the medical societies.[130]

The above-mentioned cases and views become even more instructive, if one considers them from the perspective of the Berlin philosopher Georg Simmel (1858–1918), who analysed contemporary notions of professional honour within his *Soziologie* (first published in 1908).[131] According to Simmel, honour filled the middle ground between morality and law. While morality was empowered by the inner voice of the individual's conscience, the law used external, physical force. The social demands of honour, he argued, were secured by internal as well as external means. Violations of honour had subjective, inner consequences *and* social, externally palpable ones. In this way, correct conduct could be guaranteed in areas which could not be reached by the law and in which conscience-based morality alone was not reliable.[132] In fact, the medical courts of honour punished precisely in this double manner, using personal warnings and reprimands as inner means of discipline, and employing fines, exclusion from the voting rights and publication of the verdict as social, external sanctions.

However, as Simmel realised – and as the above-mentioned cases illustrate – the problem with honour was the fact that it was tied to distinct social circles, and conflicts arose if the norms of one circle did not conform to the norms of another. Professional honour, as the honour of a social circle, made on the one hand specific extra demands and on the other hand could make allowances for certain kinds of behaviour that were even deemed dishonourable outside the circle. Simmel gave the example of the businessman, whose exaggerated praising or advertising of his goods was entirely acceptable, compared to the civil servant or the scholar in whom a similar departure from the truth would lead to a loss of honour.[133]

From such a perspective then, the problems that medical practitioners had with matters such as advertising, discounts on fees, or selling a practice, are readily understandable: they reflected conflicts between their social roles as owner of small businesses and as academics. Moreover, the strong emphasis on collegiality amongst doctors, as illustrated by the above-mentioned case from Westphalia concerning alleged sexual misconduct, can be seen as an example of conflict between specific demands of professional honour (to protect a colleague against accusations) and expectations of honourable conduct in wider society (to help an uneducated person draw up an official report). In general terms, this problem had been highlighted by the influential Leipzig professor of law, Karl Binding (1841–1920), in his inaugural address as rector of the university on the topic of honour and its vulnerability. As Binding defined it, a man acted dishonourably if he failed to fulfil his duties. For certain professions, there were special duties in addition to the general, human ones. True professional honour (*Standesehre*), according to Binding, consisted in fulfilling the special duties besides all the other, general ones, but not at the expense of the latter.[134]

A further feature of collective honour was, as Simmel observed, that the loss or violation of honour of one member of the group was felt by every other member.[135] In this sense, honour functioned as a force for social group cohesion. The medical disciplinary tribunals protected the individual doctor against insults from colleagues; but they also protected the reputation of the profession as a whole. This function of the courts of honour can be further elucidated with reference to Pierre Bourdieu's theory of cultural capital, in particular his notion of honour as symbolic capital. According to Bourdieu, individuals can accumulate, invest and transform not only economic capital (money, property) but also social capital (a person's network of 'bonds' or social relations) and cultural capital (e.g., titles and degrees from educational institutions, cultural knowledge and possessions such as books and paintings) to enhance their position in society. Honour, in Bourdieu's understanding, is a form of symbolic capital, which gives credit to its owner. Possession of honour as symbolic capital creates trust, masks or euphemises inequalities in the possession of the other three types of capital (economic, social and cultural) and stabilises power relations.[136] Building on this, one might say that a doctor's professional honour (*Standesehre*) was his personal capital as well as part of the collective cultural capital of the medical profession. A loss or violation of his honour affected therefore the trust in the individual doctor as well as in his profession, with potentially detrimental social, economic and cultural consequences. Courts of honour acted almost like banks for this symbolic capital, 'honour'. Disciplining the offender restored the collective capital of the profession and the individual capital of the offended colleague. Characteristically, this could involve a transformation of economic capital, namely the paying of a substantial fine by the offender to make amends for his violation of professional honour.[137]

The quantitative conception of honour as such was, by the way, not new with Bourdieu. Karl Binding already spoke of an individual's 'capital of honour', his or her *Ehrenkapital*.[138] A person's 'honour capital' was an anxiously guarded possession in the *Kaiserreich*. Binding referred in 1890 to the 'hysterical irritability of our sense of honour', which he regarded as 'a severe national disease', and he criticised 'the German's permanent fear that his honour may be stolen any moment by any frivolous lad, his trembling anxiety that it may already have been lost through the disdainful gesture or mocking word of a fop'.[139] Ten years later, Arthur Schnitzler's famous novella *Lieutenant Gustl* was published for the first time. A bitter satire on the officer corps of the Habsburg monarchy, it featured the inner monologue of a young lieutenant who agonises a whole night about his 'duty' to commit suicide, because he has failed to retaliate for having being insulted after a concert by a master baker – even though nobody seemed to have witnessed the incident. Only the news in the following morning that the baker has died of a stroke on the same night makes him desist from this plan – and to move on to fighting a duel over another matter in the afternoon.[140] This mindset of constantly having to guard one's honour, the excessive *Ehrgefühl*, can also be described as a specific 'habitus' in the sense of Bourdieu, that is as a permanent disposition of a person that guides their behaviour in various situations or contexts. While courts of honour controlled and enforced this behaviour, professional codes of medical conduct as well as writings on medical ethics can be seen as attempts to structure and explicate for doctors such a habitus of honour.[141]

The – from a modern perspective exaggerated – sense of honour (*Ehrgefühl*) among doctors was encapsulated in a real incident in a medical district society, which was mentioned by a speaker in the Bavarian *Landtag* in 1898: a doctor who wanted to cancel his membership of the society was challenged to a duel by the society's chairman, just for this reason.[142] Apart from illustrating the medical habitus, the force of honour in guaranteeing the cohesion of a social group, as described by Simmel, is more than obvious from this example. Courts of honour were seen by their advocates not only as necessary disciplinary structures but also as the institutional embodiment of a profession's collective honour. As a critical doctor observed in 1896, in the context of the debate about the introduction of state-authorised medical courts of honour in Prussia, the whole idea of such courts for the medical profession seemed to be based on the belief that the social respect enjoyed by military officers, civil servants, judges and lawyers was linked with their disciplinary institutions and that the medical profession would acquire a similarly high public reputation simply by following their example.[143] Looking at this statement from the perspective of Bourdieu's conception of honour, one might say that the medical courts of honour not only dealt in the symbolic capital, 'honour', but also concentrated on the social capital that was linked to membership of a particular professional group.

1.6 Conclusions

The professional ethics and decision-making of the medical disciplinary tribunals was thus deeply rooted in contemporary notions of individual as well as collective honour. Protecting or restoring honour was their ultimate raison d'être, beyond their more mundane uses to which they were also put, as instruments of professional politics. Of course, the courts of honour also had to ensure the demand for 'conscientious practice' (*gewissenhafte Berufsausübung*), a demand that had been made explicit, for example, in the central section 3 of the Prussian courts of honour law of 1899. In this sense, they also covered the area between morality (enforced by the individual's conscience) and honour, to use Simmel's sociological topography, and thus also protected the patient. However, this protection of the patient was not – as I will show in Chapter 3 – a reflection of respect for patients' interests, but an expression of medical paternalism. In the end, the doctor decided what was best for his patients, on the basis of his conscience *and* in line with current views of professional honour.

The medical disciplinary jurisdiction thus carried with it a doctor-centred ethic, which relied on notions of professional status, collective honour, personal conscience and medical paternalism. The modelling of medical courts of honour on the disciplinary institutions in the legal profession and in the military also implied the wish for a greater closeness to the state (similar to the position of these two professions), and thus for greater social authority. This need for social authority was particularly felt, as the health insurance system developed during the period of the Second Reich, with regards to insurance boards which were close to the Social Democratic movement and had considerable powers to negotiate the terms of contracts and fees with medical practitioners, thus challenging doctors' traditional independence.

2
Medical Confidentiality: The Debate on Private versus Public Interests

2.1 Introduction

Keeping the details of patients' illnesses and of their personal circumstances secret has been an ethical demand on medical practitioners since the Hippocratic Oath (c. 400 BC).[1] This demand was so frequently repeated in the literature on medical duties that it became constitutive of the doctor's ideal 'habitus' and self-image (see Chapter 4). From the modern period, it was also endorsed by legislation. The Prussian Medical Edict of 1725, for example, ruled that 'medical men must not reveal to anyone the secret faults and ailments that have been disclosed to them'. In 1794, the Prussian General Common Law extended and further detailed this demand by requiring that 'doctors, surgeons and midwives must not reveal ailments and family secrets that have come to their knowledge, except in case of a crime'. Violations of this regulation carried a fine of 5 to 50 Thaler. A specific exception was made for doctors in so far as they were explicitly obliged to report a planned crime which could not be prevented without help of the authorities.[2]

Particularly influential for further legal development in this matter became the French *Code pénal* of 1810. Its article 378 made doctors, surgeons, officers of health, pharmacists, midwives and all other persons who obtained secret information through their profession, punishable with imprisonment from one to six months or a fine of 100 to 500 Francs, if they disclosed this information – unless they were legally obliged to do so. These legal requirements of disclosure pertained to situations in which someone witnessed an attack on public security or on the life and property of an individual or had learned about a planned crime against state security. However, the regulations on these exceptions were repealed in 1832, thus making the French *secret médical* absolute.[3]

The Prussian Penal Code of 1851 largely adopted the French law on secrecy but kept the exception of authorised disclosure. According to section 155, 'medical persons and their helpers', and all other persons, were punishable with imprisonment up to three months or a fine of up to 500 Thaler if they revealed 'without authorisation' private secrets which had been entrusted to them 'through their office, profession or trade'. Some other German states, such as Hanover (1840), Hesse (1841) and Nassau (1849), had been less strict, making breaches of secrecy only punishable if they were motivated by an intention to cause harm or to gain unlawful advantages.[4] However, it was the Prussian regulation that provided the model for the section on confidentiality in the Penal Code of 1871 for the German Reich. There had been some concerns voiced by the Berlin Medical Society and the medical left-liberal member of parliament, Wilhelm Löwe (1814–1886), that too strict an interpretation of the term 'without authorisation' (*unbefugt*) might allow harassment of doctors. However, their proposals to include clauses that would make only harmful or 'improper' disclosure punishable were dismissed as unnecessary during the deliberations of the draft law in the Reichstag.[5] The approved version of section 300 of the Reich Penal Code (*Reichs-Strafgesetzbuch*) ruled

> Lawyers, advocates, notaries, counsels for the defence, physicians, surgeons, midwives, apothecaries, as well as the assistants of these persons, are punished with a fine of up to 1,500 Mark or up to three months' imprisonment, if they reveal without authorisation private secrets that have been entrusted to them through their office, profession or trade. Prosecution follows only if a petition has been filed.[6]

In view of this codification of secrecy for medical personnel, as well as the legal profession, one might assume that confidentiality would have posed no major issues in Germany after 1871. However, as my research into the medical as well as legal literature on this topic has shown, the opposite holds true. The limits of medical confidentiality, and the rationale for allowing disclosure under certain circumstances, were important issues for both the medical and the legal profession in Imperial Germany.[7] The question of medical secrecy reflected a more general ethical conflict over claims for priority of public interests versus private interests. This chapter analyses the medico-legal and moral debate on this problem. Rather than providing a strict chronology of the contemporary discussions, I will examine the main contentious issues: medical secrecy at court; warning of contact persons and notification of cases of contagious, especially venereal disease (VD); confidentiality in the context of psychiatric care; and the reporting of criminal abortion. Already in 1902, just over thirty years after section 300 had come into effect, a critical medical commentator claimed that the principle of medical confidentiality had been 'perforated in so many places that soon nothing, nothing at all, will be left of it'.[8] And another three decades on, towards the end of the Weimar Republic, a legal author stated '... in German law there exists a situation in which the duty of secrecy is, due to the many reasons permitting disclosure, merely a sieve-like scaffolding'.[9] How can these negative assessments be explained?

2.2 The doctor's right to refuse giving evidence in court

Unlike lawyers and priests, doctors had initially no right to refuse disclosure of entrusted private secrets when they were called to act as witnesses in court.[10] This omission was corrected only in 1877, through section 52 of the Code of Criminal Procedure (*Strafprozessordnung*) and section 348 of the Code of Civil Procedure (*Zivilprozessordnung*), which entitled doctors to refuse giving evidence about confidential patient information. These regulations came into effect in 1879. However, this new entitlement to remain silent in court was soon put to the test. In a divorce trial at the District Court of Frankfurt/Main, in 1884, the responsible physician and the administrator of the Rochus Hospital were called as witnesses regarding the wife's claim that her husband had been adulterous and treated in the hospital for VD. The doctor as well as the administrator refused to give evidence on the basis of their duty of confidentiality. Only after long deliberations did the court accept that they were both entitled to this refusal through the Code of Civil Procedure. The issue was particularly controversial in the case of the administrator, but the court eventually concluded that he was holding an office in which he necessarily received confidential patient information and that he was therefore likewise bound to professional secrecy.[11] It was also unclear whether female doctors practising in Germany, who had qualified abroad and were at this time not (yet) entitled to a German doctor's license (*Approbation*), had the same right to refuse giving evidence in court as licensed male doctors. A decision of the District Court of Frankfurt/Main in 1901 denied them this right.[12]

Equally controversial was (and remained so) the question whether a doctor who decided *not* to use his entitlement to refuse giving evidence in court, and thus disclosed personal patient details, might be punishable under section 300. Legal experts failed to agree on this issue. Some argued that it had not been the intention of the law to give doctors a personal choice whether they wished to disclose or not, but simply to enable them to refuse to speak so that they would not come into conflict with their legal duty of secrecy according to section 300.[13] Others maintained that a witness statement requested by a judge could never be an illegal act, and that doctors were allowed to use their discretion to decide case by case and according to their own dutiful judgement whether they should speak or remain silent. This latter view was endorsed in a Supreme Court (*Reichsgericht*) decision in 1889 and it became the dominant legal opinion.[14] Commentators from the medical profession held this opinion too, although the preserving of an ethical obligation to maintain confidentiality was emphasised.[15] However, general agreement on this issue was not reached throughout the period of the *Kaiserreich* (and neither in the Weimar Republic). Disclosing confidential patient information in court remained risky for doctors, and the expert literature accordingly advised them to exercise caution and to consider carefully whether to speak in court if the patient had not released them from their duty of confidentiality.[16]

However, the difficulties for a doctor who steadfastly refused to give evidence became apparent, for example, in a prominent Hamburg divorce trial at the start of the twentieth century. The doctor concerned had repeatedly refused to confirm to the Higher District Court of Hamburg that he had treated the husband for freshly acquired syphilis in May 1899. Such confirmation would have supported the wife's accusation that the husband had committed adultery and would therefore have facilitated the divorce for which she had filed a petition. Referring to section 300 of the Penal Code, the Code of Civil Procedure, and his personal promise of confidentiality to the husband, the doctor refused to give evidence and even stood firm when a (later rescinded) interim judgement obliged the husband to release the doctor of his duty of secrecy. The Supreme Court had to deal with the case twice and in 1903 eventually declared that the doctor was justified in refusing to give the demanded information. At this point in time, the Supreme Court reasoned, the doctor's confirmation of the husband's syphilis, back in 1899, would merely serve to lead to the desired divorce, not to protect the health of the wife. Facilitating a divorce, however, was in the court's opinion no 'higher moral duty' than the duty of medical confidentiality.[17]

While these legal decisions in divorce trials had thus eventually protected the doctor's right to refuse giving evidence, medical secrecy was seriously challenged and overridden in different contexts in which public security was (allegedly) at stake. According to section 139 of the Reich Penal Code, a person was punishable with imprisonment if he or she had obtained knowledge of a planned crime constituting a public danger and had failed to inform the police or warn the endangered persons. This duty applied to everyone – there were no exceptions for certain professions. This general duty to inform or warn was also expressed in the Law on Explosives of 1884 regarding knowledge of a planned bomb attack.[18] These two legal exceptions to medical secrecy had a clearly preventative character, that is, the duty of disclosure was meant to help prevent a dangerous crime, not to help prosecute a committed crime. While these exceptions were largely uncontroversial,[19] a public debate arose in 1903 when the police confiscated the medical casebook of a doctor in order to procure evidence for the prosecution.

In June of that year, on the evening of the Parliamentary elections, street riots had occurred in Laurahütte, a town with many Polish inhabitants in Upper Silesia. Shops were destroyed and the town hall attacked. The police intervened, using firearms – one rioter was killed and several were wounded. Authorised by the investigating judge's order, three police detectives visited a local Polish doctor the next day and demanded that he hand over his casebook, so that they could identify wounded rioters whom he might have treated. When the doctor refused, the book was removed from his office and taken

to the judge. From the information in the book two rioters were identified and immediately arrested. Unsurprisingly, this incident led to
harsh reactions from the medical profession. The Berlin correspondent
of the *Lancet*, for example, commented that the 'rights and privileges
of the medical profession' had been 'encroached on in an almost
incredible way by a judicial authority' and that the judge's course of
action was 'objectionable from an ethical point of view and prejudi-
cial to the interests of the profession as well as of the patients'.[20]

At the session of the Reichstag on 1 March 1904, a legally trained
member of parliament, Dr Ablaß of the Progressive Democratic Party
(*Fortschrittliche Volkspartei*), argued that the doctors' right to refuse
giving evidence in court, as guaranteed by the Code of Criminal
Procedure, should be interpreted as protecting the confidentiality of
medical case notes as well. He quoted with approval the Tübingen
professor of law, Ernst Beling (1866–1932), who had compared the
state authority's action in the Laurahütte case with that of a robber
who generously abstains from blackmail but simply takes by force
from the victim what he wants.[21] The prevailing legal opinion was,
however, that the judge's action in Laurahütte had been correct.
According to sections 94 and 95 of the Code of Criminal Procedure,
case notes could be *confiscated* as evidence. A doctor, being bound to
professional secrecy by section 300, just could not be *forced to hand
them over*.[22] The State Secretary in the Reich Justice Office, Arnold
Nieberding (1838–1912), accordingly pointed out in his reply to
Dr Ablaß that in current legal practice the Code of Criminal Procedure
merely protected the 'personal opinions' of a doctor against disclosure in court, but not his documents. He admitted, however, that the
matter was controversial and that it might be discussed in the context of a future penal law reform (which he had initiated in 1902).[23]

The same issue came up again, however, in 1910. In the wake of
riots in Berlin-Moabit[24] the police confiscated medical case notes
after the doctors concerned had refused to provide any information
about injured persons they might have treated. Again, legal experts
pointed out that the police's course of action was legitimate. Albert
Hellwig, a Berlin lawyer, argued that the state had a legitimate claim
to prosecuting the rioters. Doctors had to respect this higher interest
of the state, because they were in the first instance citizens with duties
to the community as a whole, not only to their individual patients.[25]
Ludwig Ebermayer (1858–1933), lawyer at the Supreme Court, referred

to this court's decision of 1903 (in the divorce case discussed above), which had set up the important principle that a 'higher moral duty' (*höhere sittliche Pflicht*) might supersede the legal duty of professional secrecy. Applying this principle, he identified the protection of the security of the state and the public good as a higher moral duty and concluded that the doctors concerned should have refrained from their right to refuse giving evidence and should have thus disclosed the required information to the police.[26] As both commentators made clear, in such cases public interest had to come before the private interest of patients.

The Supreme Court's decision of 1903 was also important for the controversial question whether potential sexual contacts of VD patients should be warned in order to prevent the spread of the infection.[27] As the court had stated

> ... as there are *legal duties*, which can have priority over the duty of secrecy (e.g. the duty to report according to section 139 ...), so there are also *higher moral* duties to be recognised, in view of which the obligation of confidentiality has to take second place. Therefore it *may* for example under certain circumstances appear to be a doctor's duty to inform the wife about her husband's venereal disease, in order to prevent her infection as far as possible; likewise, it might not be entirely ruled out to assume that such a moral duty of information may exist vis-à-vis a third person who may not be the wife.[28]

Although this opinion was very cautiously formulated and hypothetical, the possibility admitted through it, to warn sexual contacts of a patient with VD against the patient's will and without violation of section 300, became highly significant. The wider background to this opinion and its consequences will therefore be discussed in the following section.

2.3 Confidentiality and venereal diseases

The problem as such was not new. One of the standard scenarios discussed in the literature on confidentiality and medical ethics was that of the fiancé who suffers from syphilis or gonorrhoea but refuses to disclose his illness to his future wife. Doctors were usually reluctant to sacrifice medical secrecy and to break the law in this situation but felt that they should do something to prevent infection. One type of advice was to recommend to the fiancée's family that it was generally a good idea for the future husband to take out life insurance. If the latter then refused to undergo the necessary health examination for the insurance, the fiancée and her family could draw their own conclusions.[29] Another tactic was to convince a syphilitic patient to delay the date of marriage – with the expectation that the symptoms might become more visible over time, thus making disclosure by the doctor superfluous.[30] The issue found also its way into fiction and poetry. In 1902, Ottilie Franzos, writing under her pen name F. Ottmer, published her novella *Das Schweigen*, in which a doctor decides not to inform a young woman of her fiancé's infection, believing that his professional honour obliges him to maintain silence. His decision has disastrous consequences for the health of the woman and the couple's baby, who dies shortly after the father has succumbed to his illness.[31] The plot of this work later served as the blueprint for two films by the director and sex educator Richard Oswald (1880–1963), *Es werde Licht!* (Let there be light!) in 1917 and *Dürfen wir schweigen?* (May we remain silent?) in 1926.[32]

A broader context for the problem of warning contact persons was provided by the contemporary debate on the notification of infectious diseases. In principle, a duty for medical personnel, as well as for heads of households and landlords, to report dangerous contagious diseases to the police had already been introduced with the Prussian Sanitary Regulations of 1835. Regarding syphilitic infections, however, these regulations only required doctors to inform the local police authorities if keeping the disease secret was feared to have, in their medical opinion, 'negative consequences for the patient concerned or the community'. If a doctor neglected this requirement in such a case, he was punishable with a fine. In addition, doctors had the duty to report soldiers with VD to the relevant military authorities. Notification of VD in prostitutes remained compulsory on the basis of the Prussian General Common Law.[33]

Doctors seem, however, to have been reluctant to strictly follow these sanitary regulations, partly because breaches of patient confidentiality would have damaged their reputation and thus harmed their medical practice. By the end of the nineteenth century, the 1835 regulations had been virtually 'forgotten' among doctors, so that a ministerial decree in 1898 had to draw attention to the fact that they were actually still valid.[34] The authorities realised that new legislation on the notification of contagious diseases was necessary from the point of view of public health, especially after the experience of the 1892 cholera outbreak in Hamburg.[35] In 1900, the Reichstag passed a law on the 'combat of diseases constituting a danger to the public' (*Gesetz, betreffend die Bekämpfung gemeingefährlicher Krankheiten*). It included an obligation for doctors to report immediately to the police any case of suspected illness or death from leprosy, cholera, typhus, yellow fever, plague or smallpox. Significantly, syphilis and other VDs were omitted from this list.[36] The regulations of the various German states for the implementation of this Reich law extended the list of notifiable diseases. The Prussian implementation law of August 1905 included notification for cases of diphtheria, meningitis, puerperal fever, trachoma, recurrent fever, dysentery, scarlet fever, typhoid fever, rabies, trichinosis, meat and fish poisoning and – for cases of death only – pulmonary and laryngeal tuberculosis, but VDs were again left out. The draft of this implementation law had been discussed extensively in the lower house (*Abgeordnetenhaus*) of the Prussian Parliament, but the debate had been dominated by considerations of state and municipality costs, for example through compensation payments for loss of earnings of isolated individuals. Some concern had been expressed how doctors could identify in each given case whether a patient with VD was a prostitute in order to decide whether notification was required. Other concerns were expressed about the reporting of soldiers with VD to the military authorities and about the breach of section 300 of the Penal Code that this might involve.[37] In the end, however, VDs remained notifiable in only those cases that had already been specified in the Prussian Sanitary Regulations, that is essentially in soldiers and in prostitutes. In particular, prostitutes diagnosed with syphilis, gonorrhoea or chancre could be subjected to surveillance, isolation and compulsory treatment.[38]

Against this background, lively discussions about proposals to extend the general legal duty of notification of dangerous contagious diseases to VD, in the interest of public health, took place in the early years of the twentieth century. Doctors who argued against notification pointed out that disclosure of tabooed diseases such as VD was unacceptable to the general public and that such breaches of medical secrecy destroyed the relationship of trust between doctor and patient. Hence patients would keep their illness secret and turn to quacks (*Kurpfuscher*) for help, instead of seeing a qualified doctor. Critics of compulsory notification also held the view that isolation of patients with long-term conditions such as syphilis was impracticable as a general measure and that existing legislation provided enough powers against VD patients who deliberately or negligently spread their infection.[39] Partly, these arguments were thus pragmatic. But it was also apparent that doctors were keen to protect medical secrecy as a professional privilege which distinguished them from non-licensed healers and which put them on a par in this regard with the respected legal profession and the clergy.

The debate about introducing a more general duty of notification for VDs was an eminently public one. For example, the German Society for the Combat of Venereal Diseases (founded in 1902), a pressure-group whose membership ranged from bourgeois social reformers, doctors, lawyers and civil servants to social democrats, feminists and moral reform campaigners,[40] organised its second congress in Munich in March 1905, choosing as one of its main themes 'Medical Secrecy and Venereal Diseases'. The relevant session comprised papers by the Breslau professor of dermatology and discoverer of the gonorrhoea bacillus, Albert Neisser (1855–1916), the Munich lawyer *Justizrat* Max Bernstein, and the Frankfurt physician and police surgeon Max Flesch (1852–1942 or 1944). Neisser's paper was presented at the meeting by a Breslau colleague, Dr Martin Chotzen (died 1921), as Neisser was at that time carrying out syphilis research on apes on the island of Java.

Neisser took issue with the view, often expressed by medics as well as non-medics, that making VDs generally notifiable would undermine patients' trust in doctors, and thus keep them away from proper medical treatment and drive them instead into the hands of quacks. In his opinion, this argument was not convincing. A *right* of doctors to report a case of VD to the sanitary authorities (as opposed to a *duty* of doing so) would have a salutary educational effect on 'careless and frivolous' patients, that is they would know that they could avoid notification if they complied fully with the doctor's prescribed treatment and rules of conduct. And those patients who were 'conscientious' would not be put off by the doctor's right to notification, as he would not have to use it anyway in their case. Neisser knew that such a right already existed in principle through the Prussian Sanitary Regulations of 1835 and that it had not been used very much. But he was optimistic that through the increased awareness of the VD problem among doctors and the public, the right of notification would be more effective, especially if doctors reported to a medical, sanitary commission, not directly to the police. Accordingly, Neisser suggested an amendment to section 300 that would give doctors such a right to report in order to prevent infection and harm to other persons. On the basis of these considerations he also proposed that in court doctors could be forced by the judge to divulge patient information if this was regarded as essential for deciding on the case. As Neisser made clear, the interests of the public had to have priority 'in all circumstances' over the interests of the individual.[41]

The lawyer Bernstein, in contrast, saw no need to amend section 300 in this way. There already existed, in his view, sufficient exceptions from the legal duty of medical confidentiality: through the patient's release of the doctor from his professional secrecy; through defined legal requirements of disclosure, as in section 139 and the recent Law on the Combat of Diseases Constituting a Danger to the Public; and, importantly, through the Supreme Court decision of 1903 that 'higher moral duties' demanding disclosure could take precedence over medical secrecy. Unlike Neisser, Bernstein wanted to leave the decision to the doctor whether he wanted to speak or remain silent in court about his patient's details. If the patient did not release the doctor from the duty of secrecy and the doctor chose to maintain silence, the judge was also entitled to draw his conclusions from these circumstances.[42]

An even more radical proposal than Neisser's was made, however, by Max Flesch. He argued that effective action against the spread of venereal infections required a general duty (not just a right) of doctors to report all VD cases to the sanitary authorities, as with the other notifiable diseases falling under the law of 1900. In his opinion, a breach of confidentiality would not exist if the civil servants of sanitary authorities, as well as the employees of health insurance organisations, were bound to professional secrecy. On this basis he advocated measures such as compulsory hospitalisation of renitent patients and, in certain cases, reporting them to the courts (which would also be bound to confidentiality). In order to prevent VD patients switching to treatment by quacks, he suggested a duty of notification also for lay healers.[43]

These papers at the Munich congress, which were subsequently hotly debated,[44] illustrated the wide spectrum of opinions on the conflict between medical confidentiality and strategies to prevent the spread of VDs. A majority of discussants, medical and non-medical, expressed however a reluctance to weaken section 300 by including Flesch's proposal of obligatory notification of VD. And, importantly, Bernstein had pointed out that Neisser's proposal of introducing a medical *right* to notification of VD might actually be seen as superfluous in light of the recognition of 'higher moral duties' in the Supreme Court decision of 1903. In fact, while these matters were intensely debated, an important test case went through the courts. Given the significance of this case, some of its details need to be given here.

A Berlin medical practitioner, Dr L., had warned a married woman, Mrs I., that her children were in danger of syphilis infection by her sister-in-law, who lived in the same household and sometimes took the children into bed with her. Mrs I. disclosed her conversation with the doctor to a neighbouring tenant, who subsequently spread the news about the sister-in-law's syphilis all over the house. When the sister-in-law and her mother then went together to Dr L. and the mother asked him what rumours he was spreading, the doctor frankly declared that her daughter had syphilis, and a heated argument followed.[45]

In February 1905, the I. District Court of Berlin found Dr L. guilty of unauthorised disclosure of patient details on two counts, that is disclosure to Mrs. I. and to the mother of the sister-in-law, and sentenced him to a fine of 20 Reich Mark according to section 300. Mitigating circumstances had been taken into account, as it was recognised that the doctor had breached confidentiality in the interest of the children in order to warn their mother of an actual existing danger of VD infection.[46]

The matter might have rested with this rather mild punishment, but Dr L. appealed against the verdict. Albert Moll, who had just published his handbook on medical ethics (*Ärztliche Ethik*, 1902), had persuaded him to challenge the verdict and acted as an expert adviser for the defence.[47] The case went up before the Supreme Court in Leipzig, which issued a significant decision on 16 May 1905. The court found that the doctor had been authorised to disclose the sister-in-law's syphilis to the children's mother through his duty of conscientious professional practice, as required in the Prussian Law on Medical Courts of Honour of 1899 (see Chapter 1). The duty to warn of an infection might be seen as part of this professional duty of conscientious practice. The court recognised the defendant's argument that he had regarded it as his duty to warn the children's mother. It also accepted his claim (which seems to have been derived from a suggestion made by Moll) that he had feared being prosecuted for having caused bodily harm through negligence if he failed to warn. In fact, such fears may not have been entirely unfounded, because deliberate or negligent infection of a person with VD was punishable and the offender liable to pay damages under the physical injury legislation.[48] Moreover, Dr L. was exonerated from unauthorised disclosure vis-à-vis the sister-in-law's mother, because at that time her daughter's illness had become public knowledge in the neighbourhood. Also, since the sister-in-law had come together with her mother to see the doctor, it could be presumed that she tacitly consented to her mother being informed.[49] The case was referred back to the level of the lower court, and the II. Berlin District Court duly acquitted the doctor.[50]

The Supreme Court had thus further pursued the line that it had taken in the divorce case of 1903 when it recognised 'higher moral duties' which might justify a breach of medical confidentiality. Obviously, the Berlin case of 1905 had been used as a test case for this rationale.[51] The Supreme Court's decision reflected the strength of popular sentiment that the duty of professional secrecy should be overridden in the fight against VD, and accordingly it attracted much attention in medical as well as legal circles. While it was welcomed among medics, jurists had divided opinions on it. Legally, as was pointed out, the court's decision stood on weak ground. It was seen as highly problematic to give *professional* duties, stated in a law of one German state (i.e., Prussia), priority over a section in the Penal Code for the whole Reich.[52] It was felt that it meant embarking on a slippery slope towards arbitrariness if decision-making on disclosure was left to the individual doctor. Moreover, exonerating Dr L. from his second breach of confidence (in speaking to his patient's mother) stood partly in contradiction to an earlier decision of the Supreme Court. In this decision, in 1894, the court had endorsed the punishment of a doctor who had confirmed a rumour about a patient's condition.[53] Others, however, defended the Supreme Court as being in tune with current conceptions of legality. The doctor should strike a balance, in each individual case, between the 'material and ethical harm' *caused* by disclosure and any such harm *prevented* by disclosure.[54]

In March 1906, the Supreme Court's controversial decision in the Berlin case became the subject of debate in the upper house (*Herrenhaus*) of the Prussian Parliament. In the context of the implementation regulations for the Prussian law of 1905 on the combat of contagious diseases, Count von Hutten-Czapski characterised the recent court decision as one that was 'apt to undermine patients' trust in their doctor and make them turn to quacks'. Accordingly, he asked the Minister for Medical Affairs to make sure that any such future cases of breaches of professional secrecy were subjected to the disciplinary proceedings of the medical courts of honour.[55] Replying for the government, Adolph Förster, the director of the ministry's medical department, shared von Hutten-Czapski's concerns that this Supreme Court decision could and would in future prevent some patients from seeing a doctor on discreet matters. In his view, the doctor–patient relationship, being a contract relationship based on trust and confidentiality, had been 'seriously shaken' by this decision, which placed the choice between disclosure and secrecy at the doctor's personal discretion. While Förster acknowledged the public health argument in favour of disclosure, he hoped that the Supreme Court would soon correct its position in this matter and that penal law reform might find a better balance between public interest and the relationship of trust between doctor and patient.[56]

However, the Supreme Court did not change its opinion on this question, and Förster himself, in his capacity as chairman of the Prussian Court of Honour for Doctors (see Chapter 1) soon followed the new line. In September 1907, this court of honour acquitted a doctor who had breached the confidence of a patient suffering from syphilis by reporting her and her sexual partner, a teacher, to the local inspector of schools. Like the Supreme Court in the case of Dr L., the court of honour acknowledged that the doctor had experienced a collision of duties – between the duty of professional secrecy towards his patient and the duty to warn of a danger to public health, that is, the health of the schoolchildren. The doctor's decision to warn of the danger of syphilitic infection was seen as justified under the given circumstances, and the court confirmed that his conduct had not violated the honour and reputation of the profession.[57]

Collectively, the two Supreme Court decisions of 1903 and 1905 and the verdict of the Prussian Court of Honour for Doctors in September 1907 established a new interpretation of 'authorised' disclosure, which could now extend to justifications of breaches of confidentiality in the name of public health or in the interest of the health of third parties. It was even suggested in the above-mentioned debate of the Prussian Parliament to amend section 300 by including a clause that would generally exempt disclosure from punishment if it were made for the protection of 'public interests'.[58] These interests of the community were increasingly seen as superior to the individual's interest in privacy. As discussed above, the notion of higher interests was also applied to justify the state's access to medical case notes in order to aid criminal prosecution. By 1910, the law on medical secrecy had been undermined through the interests of public health and justice. This development was noted with concern by parts of the medical profession. For example, when during the First World War, advice centres for VD patients were established by the social insurance organisations (*Landesversicherungsanstalten*)[59] and it was suggested that doctors should report VD patients, by name and via their health insurance organisation, to these centres, fears were expressed that the duty of medical secrecy would be seriously compromised. As an editorial in the *Ärztliches Vereinsblatt für Deutschland* complained in 1916: 'Nowadays one uses every opportunity to put forward the higher moral duty of public health, to which the duty of secrecy is supposed to subordinate itself.'[60]

The complexities created through such developments were reflected in the growth of specialist literature on the doctor's duty of secrecy, which also included increasing details about disclosure to health and life insurances and when issuing medical certificates. The volume of the medical standard text on professional secrecy, *Das Berufsgeheimnis des Arztes* by the Berlin psychiatrist Siegfried Placzek (1866–1946), increased from a mere 77 pages in the first edition of 1893 to a 230-page book in the third edition of 1909.[61] Apart from numerous articles in medical as well as legal journals, several law theses examined the topic in detail. During the first decade of the twentieth century, doctoral theses on the issue of medical secrecy were completed at the law faculties of Rostock (1903), Leipzig (1903, 1907), Freiburg (1906), Heidelberg (1907, 1909) and Breslau (1909).[62] In addition, the Heidelberg law faculty awarded an essay prize on the topic in 1906.[63]

Partly, these legal analyses were undertaken in view of the planned reform of the Reich Penal Code. In fact, a substantial legal mono-graph on section 300 was published in 1910, just after the Preliminary Draft of a New German Penal Code (1909) had been released.[64] The newly drafted section (section 268) on professional secrecy broad-ened the coverage of legal and medical personnel, extending the duty of confidentiality among others to the nursing and clerical staff of public and private hospitals. It also protected patient information that had become 'accessible' through professional practice (not just 'entrusted' by the patient), taking account of a wider interpretation that had already been established through a Supreme Court decision in 1885.[65] However, the new draft section did not include the per-sonnel of health and life insurance organisations (this was left to future insurance legislation) and neither was the duty of confidenti-ality extended to non-licensed healers (i.e., the so-called *Kurpfuscher*). Importantly, the draft kept the formulation of 'unauthorised' (*unbe-fugte*) disclosure, which left the implication of 'authorised' types of disclosure undefined.[66] Even if the penal reform had been successful (which it was not[67]), the status of professional secrecy would have remained controversial and precarious. Moreover, apart from the scenarios discussed above, there were other areas in which medical confidentiality was under threat: psychiatric care and criminal abor-tion. I will turn to these areas in the following section.

2.4 Psychiatric care and criminal abortion

As psychiatrists pointed out, the very process of admitting a patient to a mental asylum undermined their duty of secrecy. The patient's birth certificate, certificates from the patient's local authorities on maintenance rights and obligations, a medical assessment of the existence of a mental disturbance and permission from the police had to be obtained, thus involving numerous persons. While it could be assumed that individuals involved in these bureaucratic processes were sworn to maintain confidentiality, it could hardly be said that the patient's mental illness was kept secret. Moreover, public as well as private mental asylums were obliged to provide information on their patients to the public prosecutor and the courts. Particularly in proceedings for the certification of a mental patient this duty could become embarrassing for the psychiatrist, because it was a legal requirement that the court's decision was posted to the patient together with the text of the medical assessment. Finally, section 139 of the Penal Code which required the reporting of preventable dangerous crimes was often applicable in psychiatric practice. If patients with a remaining risk of self-harm that could involve harm to others (e.g., suicide by setting oneself alight, running amok) were dismissed from the hospital, the relevant police authorities had to be informed.[68]

While psychiatric commentators reluctantly accepted these required disclosures, they felt under pressure from relatives of their patients. If the latter suffered from progressive paralysis (i.e., an advanced stage of syphilis), a frequent condition in asylums around 1900, the problem of disclosure in order to warn a prospective bride or father-in-law was felt particularly strongly. The Supreme Court decision of 1903, with its stipulation of 'higher moral duties', was taken into account, but there was still reluctance to sacrifice the patient's interest in his privacy. Similar issues arose in cases of psychiatric conditions that were deemed hereditary. Moreover, medical statements were often required by relatives about a deceased patient's mental condition during the time when he made his last will. Although psychiatrists felt obliged to protect the privacy of their former patient, they were inclined to make such statements in order to prevent injustices. However, they were on guard vis-à-vis requests from life insurance companies, especially if they were made over the phone.[69]

Finally, the context of academic psychiatry[70] posed specific risks for confidentiality. Clinical lectures demonstrating psychiatric patients not only attracted medical students or doctors but also interested lay persons and the press. The general advice in these cases was to ensure the prior consent of the relatives and of the patient (even if the latter's consent might not be legally valid). The patient's refusal to take part in the demonstration should be accepted from a therapeutic perspective. Consent of the relatives and the patient was also advised when publishing psychiatric case histories. Names, residence and other details of the patient that made him identifiable were supposed to be omitted. There was also concern about reproducing patient photographs in psychiatric textbooks. Such photographs had become increasingly fashionable in the early twentieth century, although their didactic value seemed doubtful.[71]

In general, psychiatrists' comments on confidentiality reflected considerable sensitivity on this issue and an effort to preserve the privilege of secrecy – in the interest of their profession as well as that of their patients. Such sensitivity may well have to be seen against the background of an 'anti-psychiatric' lay movement in Imperial Germany, which accused psychiatrists of abuse of their powers, especially in the process of compulsory hospitalisation.[72]

A defensive attitude among doctors also became apparent in the problematic issue of reporting cases of criminal abortion. In principle, the Prussian Sanitary Regulations of 1835 obliged medical personnel, as well as 'all family members, householders and landlords' to report without delay any 'sudden, suspicious illnesses or deaths' to the police. This could be, and was, interpreted as including abortion attempts which had led to complications and thus came to the attention of doctors. In Baden, a ministerial decree of 1883 directly required doctors to report cases of criminal abortion.[73] Under section 218 of the Reich Penal Code of 1871 abortion was punishable by 6 months' imprisonment up to 5 years of penal servitude. Although they were also illegal in principle, this did not apply to medically indicated abortions carried out by doctors in order to save the mother's life – typically when a too narrow pelvis prevented a natural birth. Caesarean sections still had a very high mortality rate for mother and child. State guidelines declared abortion performed by a doctor as justified if it was the only option to remove a grave danger to the health or life of the mother. A Supreme Court decision explicitly confirming the legality of therapeutic abortion was issued in 1927.[74]

In the earlier medical deontological literature denunciation of abortion attempts was occasionally advocated,[75] but in the late nineteenth and early twentieth centuries doctors had become protective of the pregnant women concerned (see Chapter 4). For example, in 1894 several doctors in the county of Lennep unanimously refused to give evidence when they were called to act as witnesses in an investigation of criminal abortion. As the district physician of Lennep subsequently explained 'the girl or married woman who comes to treatment with an imminent miscarriage or septic infection', showing 'significant signs' of a criminal abortion, was protected by the professional secrecy of section 300.[76] In 1914, the Supreme Court acknowledged, in a similar case, that the medical duty of confidentiality could entitle doctors to give incomplete evidence in court, as long as this did not lead to incorrect and untrue statements. In this case two doctors had confirmed that the defendant had carried out abortions but refused to give the names of the women concerned.[77] Reporting a woman who had undergone an abortion was regarded as an unusual step for a doctor. Albert Moll commented on such a case in 1911, when a doctor had reported the woman with her consent in order to initiate prosecution of the abortionist. Both the abortionist and the woman concerned were subsequently convicted, but the latter was able to submit a plea for clemency which was supported by the public prosecutor. In Moll's opinion, a case like this constituted a matter for personal conscientious decision-making of the doctor, but he warned against trying to construct a 'higher moral duty' (in the sense of the Supreme Court decision of 1903) at any such occasion.[78]

The transition in medical attitudes on this issue was also reflected in the different editions of Placzek's book on medical confidentiality. In the 1893 edition, he still held that most doctors, if they noticed that a criminal abortion had taken place, would report the case to the police. He assured his readers that such disclosure to the authorities could not be punished. But simultaneously he pointed out that a doctor who did not report in this situation would act 'absolutely correctly', and that the decision to report or not to report had to be left to the 'tact and sense of duty' of the individual doctor.[79] For the third edition of his book in 1909, Placzek revised the relevant passage and now made a clearer, though still cautious, recommendation:

> The most fitting conduct of the doctor might be to report to the police only those cases in which the imminent crime may still be prevented. If the doctor merely has to note a 'fait accompli', the sole motive to see one more person punished cannot justify the report. The resulting use would be out of proportion to the seriousness of the breach of trust of which the doctor would be guilty vis-à-vis his patient.[80]

Placzek admitted that in two cases he himself had not respected his own principle, when he had been asked to provide an abortifacient and refrained from reporting these incidents to the police. He had merely rejected the requests and emphasised their illegal nature. The implication of this admission was thus to sanction professional secrecy even if an abortion might have been prevented by disclosure. Placzek was inclined to report abortionists (e.g., a midwife who has performed the abortion), but hesitated even in this situation, as the pregnant woman was likely to be exposed and punished as well. Acknowledging the social plight of women seeking an abortion because they already had many children, he was reluctant to report. Only if the pregnant woman had died from an abortion attempt was he fully prepared to report the abortionist (if known) to the police.[81]

A similar line had been taken by Placzek's colleague Albert Moll in his book of 1902 on medical ethics:

> If an author says that the doctor should, when called to an artificial miscarriage, report the woman who has induced it, then this does not merit any further discussion. From the perspective of ethics, the significance of abortion is judged very differently, and an artificial miscarriage is not generally as disapproved of as most other criminal acts. I doubt whether under these circumstances the conflict of duties is so great as to justify a violation of the duty of confidentiality, especially since the client would be exposed to a conviction. ... Just in order to have one person punished because she has committed a crime which is widely held not to constitute a danger to the public, the doctor will hardly be allowed to deviate from his duty of confidentiality.[82]

The reluctance of doctors to report abortions to the police is likely to have contributed to very low conviction rates. Placzek, who was aware of this issue, estimated that approximately 1 per cent of cases of criminal abortion ended with a conviction and punishment.[83] However, according to figures compiled in a recent law thesis, the number of convictions for criminal abortion in Germany between 1900 and 1918 rose from 314 to 1,730. During the Weimar Republic the number of convictions rose temporarily to 7,193 (in 1925), no doubt reflecting significant increases in abortion rates due to the economic crises, and was at 3,809 in 1933, the year the National Socialist regime seized power. These statistics do not, however, include figures on the annual number of cases that were reported to the police.[84] It is probably safe to assume that the growing number of convictions reflected an increase in the total number of abortions rather than any significant change in doctors' attitudes towards reporting such cases. Although, as Cornelie Usborne has pointed out, the medical profession took in general a pro-natalist, anti-abortion line in public discussions about abortion law reform before the First World War, the evidence on doctors' attitude towards reporting women who had made an abortion attempt indicates sympathy with women in conflict situations. Calls of the women's movement to repeal or change section 218 had become prominent in the early 1900s, and the mitigation of the penalty for an aborting woman was part of the deliberations on a general penal law reform. The Preliminary Draft of 1909 for a new penal code included a maximum penalty of three years imprisonment for a woman who had undergone an abortion. In the Weimar Republic, abortion law reform developed into a major political issue, as the left parties lent their support to the women's movement in this question.[85]

2.5 Conclusions

The view that a doctor's duties to the general public should carry more weight than his duty of confidentiality to the individual patient continued to be relevant during the Weimar Republic. It was particularly reflected in the Law for the Combat of Venereal Diseases (*Gesetz zur Bekämpfung der Geschlechtskrankheiten*) of 1927, which required that doctors reported contagious VD patients who refused treatment, or who posed a danger to the public through their occupation, to the health authorities or to the advice centres. Although the employees of the health authorities and VD advice centres were bound to secrecy through the same law, they could disclose a patient's disease to other officials or another person who had 'a justified interest' in being informed about the patient's VD.[86] The doctrine of the 'collision of duties', as formulated through the Supreme Court decision of 1905 in the Berlin case, and the concept of higher moral duties to the public had thus found their way into legislation.

The history of medical confidentiality in Germany between 1871 and 1927 was thus one of an ever-increasing limitation on the doctor's right to remain silent. Legal exceptions from medical secrecy were made in order to prevent and prosecute crimes, to prevent epidemics of dangerous contagious diseases and to fight the spread of VD. The notion of a 'higher moral duty' that could justify the breach of a patient's confidence was introduced in Supreme Court decisions in Imperial Germany and was implemented in law during the Weimar Republic. The general justification for sacrificing individual patients' interest in privacy consisted in the priority of public interests – in health and in the prosecution of crimes. In some areas, especially on the issue of illegal abortion, doctors were protective about their right to secrecy, not only in the interest of the patient concerned but also in their own interest, as medical confidentiality constituted an important professional privilege which they shared with lawyers and the clergy. Moreover, discreetness was part of medical professional honour and of the doctor's habitus (see also Chapter 4). However, doctors were also open to public health and political arguments that spoke for restricting their right to remain silent. In the end, they failed to resist the legal and political trends that increasingly perforated medical confidentiality until it became a 'sieve-like' structure. In this way, doctors showed a move away from a traditional ethos that focused on their relationship to the individual patient towards a collective ethics that put the (presumed) interests of society and of the state first.

3
Patient Information and Consent: Self-Determination versus Paternalism

3.1 Introduction

One of the remarkable features of medical ethics in Imperial Germany was its ambiguous and sometimes openly controversial relationship with the law. An area where a conflict between doctors' and lawyers' differing conceptions of the therapeutic relationship was played out was the issue of patient consent to medical interventions, in particular to surgical operations. This chapter traces the debate on this issue from its beginnings in the early 1890s to the First World War. As I will show, the traditional paternalism of physicians and surgeons was challenged in court decisions, which forced them to adopt practices of consent-seeking and to consider the appropriate amount of patient information before treatment. However, in spite of those new requirements, doctors kept a paternalistic attitude.

The debate on patient information and consent started with a controversy among legal experts in Switzerland. On 17 June 1892, Lassa Oppenheim (1858–1919), newly appointed professor of criminal law at the University of Basle, gave his public inaugural lecture on the topic 'The Medical Right to Bodily Interventions on Sick and Healthy Persons'. Oppenheim pointed to the increase in surgical procedures after the introduction of antiseptic methods and to new therapies in internal medicine, such as the tuberculosis treatments with Robert Koch's secret remedy 'tuberculin', which had recently risen 'like a meteor in the sky of medicine' but had also resulted in the death of several patients.[1] What are the limits for doctors' interventions into their patients' bodies, asked Oppenheim; when do medical treatments collide with penal law? Moreover, how can experiments on healthy or ill persons legally be justified?[2]

3.2 Legal positions and the Oppenheim–Stooss controversy

The issue had previously been treated, though only briefly and in rather general terms, by a number of penologists. Crucial in this regard for the German Reich was the interpretation of section 223 of the Penal Code (*Reichs-Strafgesetzbuch*) of 1871, which ruled that 'someone who intentionally mistreats somebody else's body or harms that person's health, will be punished for assault and battery with imprisonment up to three years or a fine of up to one thousand Mark'. The classical justification for a bodily intervention without legal consequences had been the prior consent of the injured person, according to the principle of Roman law, *volenti non fit injuria* (i.e., a consenting person cannot be wronged).[3]

However, according to more recent conceptions of legality this simple justification was no longer tenable. Human life, health and bodily integrity were classified as inalienable goods which therefore could not be justifiably violated on the basis of consent alone. Some theorists of law, such as the Bonn professor Hugo Hälschner (1817–1889), emphasised that only the *moral purpose* of medical interventions, derived from a moral duty of patients to preserve their life, made the patient's consent to an intervention legally permissible.[4] But there were medical procedures without, or at least with dubious, moral purposes, such as those of purely cosmetic surgery, as Oppenheim pointed out.[5] Moreover, there were difficulties with a strict requirement of consent in other medical scenarios. For example, the forcible treatment of someone who had made a suicide attempt, deliberately against that person's will, was deemed acceptable medical practice. Similarly, it was regarded as inappropriate to expect medical practitioners to educate their patients about the potential side effects of a remedy and to ask them for their consent before prescribing it.[6]

Another legal view, held for example by the professors of law Karl Binding (1841–1920) in Leipzig and Franz von Liszt (1851–1919) in Marburg, was that state-licensed doctors had a professional right (*Berufsrecht*) to carry out physical interventions on their patients.

This right was thought to exist for as long as the procedures were not performed against the patient's will or, in case of unconscious or mentally ill patients, against the wishes of their relatives. However, as a basic principle this justification of a professional right was questionable for Oppenheim: not every doctor was entitled to perform any operation, but only the ones for which he had acquired the necessary skills; and a layperson that conducted an appropriate medical procedure, for example in an emergency or accident, would hardly be punished for assault and battery.[7]

Oppenheim proposed instead in his lecture that the legal basis of medical interventions lay in historical custom and practice. Doctors had acquired a right to bodily interventions through common law (*Gewohnheitsrecht*), which recognised the medical purpose (*ärztlicher Zweck*) as the justification of these procedures. Still, consent was essential. The population's sense of justice, claimed Oppenheim, required consent as a condition for justified medical interventions. In line with Binding, who had been his academic teacher, he adhered to the principle that no mentally competent adult had to tolerate medical treatment against his or her will. Any operation (except in medical emergencies, including suicide attempts) required the patient's consent. Interventions without consent constituted either physical injury (*Körperverletzung*) or at least coercion (*Nötigung*). Moreover, before any dangerous operation was carried out, the patient had to be made fully aware of the risks. Never, stressed Oppenheim, was an experiment on a person – whether healthy, sick, terminally ill or even convicted to death and awaiting execution – permissible without their consent. For example, an inoculation experiment by the Prague dermatologist Johann Ritter von Waller concerning the transmission of secondary syphilis, carried out in 1851 without information or consent on a 12-year-old boy, was described as 'dreadful' by Oppenheim, and he left no doubt that such trials had to be regarded as punishable bodily harm.[8]

To the published version of his lecture Oppenheim appended two court decisions, which in his opinion further demonstrated how necessary it was to determine the 'legal foundation and limits of the medical right to bodily interventions'.[9] In one case (from 1882), the Criminal Court of Basle City had acquitted a doctor, who had transplanted skin from a 15-year-old servant girl (with her consent) to a badly healing wound of her employer, although the girl's father had not given his approval. In the other case, dating from May 1892, the Kassel District Court had sentenced a psychiatrist to three months' imprisonment according to section 223, because he had tried to silence a hysterically screaming elderly lady in his private clinic by repeatedly beating her – with his hand, with a stick and eventually with a riding whip.[10]

The fact that Oppenheim referred in his argument to surgical, medical and psychiatric treatments as well as to scientific experiments reflected his interest in a *generic* legal assessment of all kinds of medical interventions. The initially enthusiastic reception (by doctors as well as patients) and fast clinical introduction of Koch's 'tuberculin' in 1890/91, despite the substance's severe side effects, had shown that the boundaries between therapeutic and experimental procedures were fluid. If an ultimately therapeutic intention could be claimed, issues of consent seemed less important. Even for control injections with tuberculin in healthy individuals or patients with diseases other than tuberculosis consent was rarely obtained.[11] However, Oppenheim was not really concerned with the distinction between therapeutic and non-therapeutic procedures. He even naively predicted that illegal human experimentation would come to an end once doctors had been 'educated' about the limits of their right to bodily interventions.[12]

It was indeed Oppenheim's very general, common-law approach to the question of a medical right to bodily interventions that provoked criticism in the first instance. Carl Stooss (1849–1934), professor of law in Berne, immediately replied in the *Zeitschrift für Schweizer Strafrecht* (of which he was editor), first in a brief review[13] and subsequently in a full article. Stooss thought that Oppenheim's argument of the medical purpose of an intervention as its common-law justification was wrong, because a doctor might carry out an inappropriate procedure with the best medical intentions. What was decisive for the legality of a bodily intervention was, instead, that it was necessary according to medical experience and medical science and that it was performed *lege artis*.[14]

Moreover, Stooss denied that the condition of consent was a legal rule that had been anchored in the population's consciousness and that had been practised as law for a long time; that is, he rejected Oppenheim's claim that it was part of common law. Accordingly, he took a rather ambiguous stance on the requirement of consent.

On the one hand he stated that the doctor needed to obtain the consent of the patient (or, under certain circumstances, of their legal guardian) before an operation, if seeking this consent was possible, because nobody was obliged to subject themselves to physical treatment unless there was an exception stated in the law. On the other hand he raised the question whether a doctor who had operated on a patient without their consent and had caused damage to their body – but was not guilty of anything else – was punishable from the perspective of civil law, that is whether compensation could be claimed from him. He suggested treating such a case merely as an infringement that was punishable by the police – not as a crime.[15] In addition, he argued that operations were justified by their success, that they did not harm the patient's health, and therefore could not be classified as physical injury in the sense of the law. If the result of an operation were unexpectedly unsuccessful, the doctor would have to be acquitted in a criminal court, because in penal law the expectation of success – if this expectation rested on an excusable error – was treated as equivalent to real success.[16]

Immediately Oppenheim passionately defended his view in the same journal, claiming that his common-law justification of medical interventions had been based on his extensive study of medical literature and on personal discussions with medical doctors. Even successful operations still factually remained physical injuries. Regardless of the success of an operation, regardless whether the patient's health had been damaged or not, a doctor who had operated without the patient's consent was punishable because of physical injury. 'Day by day', he maintained, patients were dying because they refused an operation that could have prolonged their lives, and surgeons knew that they would be punishable for battery if they operated in such situations.[17] Making a plea for the self-determination of patients, Oppenheim concluded

> I, for one, have no doubt that the courts, having heard medical experts, will punish a doctor who operates against the patient's will, for battery and coercion...It has to be left to the patient whether he wants to become crippled but healed, or whether he wants to stay ill or even die. ...The doctor does not yet rule over the world, it is still a matter of personal trust whether I want to subject myself to the doctor or not. And it is good that it is like that![18]

Stooss responded by insisting that successful operations could not legally be classified as physical injuries and elaborated on his criticism of Oppenheim's common-law approach.[19]

The issue of consent to medical interventions might have remained a topic of rather theoretical legal debate. Oppenheim seems to have been right in his observation that surgeons in his time generally respected their patients' wishes if they explicitly refused to undergo an operation. Karen Nolte has recently shown, from published case histories of the first half of the nineteenth century, that surgeons usually discussed very painful, dangerous or mutilating operations with their patients, regardless of their social status, and sought their consent.[20] This consent was deemed necessary to secure the patient's collaboration and to improve the chances of recovery from the operation. After the advent of effective anaesthesia in the mid-1840s, surgeons became more prone to invasive treatments, but still respected patients' refusals. For example, a look into a surgical journal article of 1889 on bone cancer (osteosarcoma), summarising a total of 44 case histories, shows that in 4 cases no operation was carried out, because the patients (or, in the case of children, their parents) had refused it.[21] In a recent MD thesis Marianne Sinn has examined surgical journals between 1894 and 1933 for reports on operation refusals and found that such cases were regularly listed, usually without further comment, in the statistics – the number of refusals varying between approximately 3 and 30 per cent of the cases.[22] It was only seen as problematic if people refused minor operations to restore their capacity to work and continued to claim incapacity benefit on the workers' accident insurance.[23] However in 1894, the year of Stooss' second response to Oppenheim, the German Supreme Court (*Reichsgericht*) in Leipzig made a landmark decision, which put the issue of consent to medical interventions on the agenda for a much wider legal, as well as medical, debate.

3.3 The 'Hamburg case'

In May 1894, the Supreme Court had to consider the case of the Hamburg surgeon Dr Heinrich Waitz. He had carried out a resection of tuberculous and purulent bones of the foot of a 7-year-old girl – against the explicit, repeatedly expressed wishes of the girl's father, a landlord, who was an adherent of non-invasive naturopathy. On the day of the operation, as the procedure was just about to start, the father, Mr K., had appeared in the hospital to take his daughter home, but Dr Waitz then decided that it was 'too late' and proceeded with the resection.[24] As medical experts confirmed in the subsequent trial for physical injury (according to section 223 of the German Penal Code), the operation was medically necessary, even life-saving, and had been performed *lege artis*. In fact, the court of first instance, the Hamburg District Court, acquitted the surgeon, explaining in its reasons that the girl's health had not been harmed, but actually been improved by the operation. The Hamburg court thus followed a similar line of argument as Stooss. The lack of consent of the girl's father, the court further stated in its reasons, did not turn this necessary medical intervention into assault and battery, and Waitz had not operated with an awareness of doing something illegal.[25]

However, the Hamburg public prosecutor, Richard Keßler (1849–1908), successfully appealed against this decision and the case went up to the Supreme Court. Keßler had particular expertise in the area of battery law and consent, having published a monograph on this subject in 1884 – several years before the Oppenheim–Stooss controversy and when he was still a district court judge in Lüneburg. Here he had argued that it was only the consent of the patient (or their legal representative), not the 'moral purpose' in the sense of Hälschner, that legally justified medical interventions and excluded them from punishment for battery.[26] It is therefore likely that Keßler regarded the case of Waitz as an important test case for his own views. The same may be assumed for the legal representative of Mr K. as joint plaintiff, the lawyer Lothar Volkmar (1852–1902), who was an active campaigner in the contemporary natural healing and life-reform movement and who publicised the trial in his own naturopathy journal, *Neue Heilkunst*.[27]

In its decision, issued on 31 May 1894, the Supreme Court endorsed the legal view that medical interventions constituted objectively physical injuries in the sense of the penal law. They were punishable under section 223 if the doctor was unable to derive his right to operate 'from an existing contractual relationship or the presumptive consent, the assumed instruction, of duly legitimised persons'. Even more so a doctor was guilty of physical injury if he acted against the declared wishes of this relevant person.[28] In other words, the Supreme Court followed the line of Oppenheim and Keßler which had made patient consent the essential condition for the legality of medical interventions, and it rejected the point of view argued by Stooss that battery law was inapplicable to surgical procedures. For the court, at least a form of tacit consent had to be demonstrable. Otherwise, the doctor's intervention was punishable as assault and battery. In the case of Dr Waitz this meant that the Supreme Court regarded him as guilty, and accordingly it lifted the verdict of the first instance and referred his case back to the Hamburg District Court for reconsideration.

Although Waitz's case ended with his acquittal on a technicality – the Hamburg court argued that he had to proceed with the operation in the given situation due to imminent danger[29] – the Supreme Court had set an important precedent on the question of patient consent. Medical paternalism had been seriously challenged. Conceptions of legality that emphasised a patient's right to self-determination, and thus the importance of their consent, had won over more traditional views that had assumed a professional right of doctors to bodily interventions. As Thorsten Noack has recently argued, this decision of the Supreme Court constituted a break with current legal theory of the time and probably owed much to one particular judge, Otto Mittelstädt (1834–1899), who was known for his liberal and rather unorthodox views.[30] How did the medical profession, and experts in law, react to this decision? What were its long-term implications for medical practice?

3.4 Medical points of view and the 'Dresden case'

An early medical response was published in August 1894 in the journal of the German Association of Doctors' Societies, the *Ärztliches Vereinsblatt für Deutschland:* Dr Justus Thiersch, a Leipzig medical practitioner who was active in professional politics, critically discussed the decision of the Supreme Court. While he agreed with the court's view that Waitz had acted illegally by operating without the father's consent, he took offence at the interpretation of surgical interventions as bodily injuries and mistreatments in the sense of the penal law. From his perspective as a medical man, such a legal view ignored the good intentions and purposes inextricably linked to medical interventions, and he therefore felt obliged 'to make a stand' against this 'completely *unjustified* interpretation of the notion of mistreatment'.[31] Thiersch's indignation was echoed in an editorial in the following month, which expressed 'astonishment' at the court's equation of a surgical operation with 'physical mistreatment'.[32] Some years later, in 1899, the Munich professor of surgery, Ottmar von Angerer (1850–1918), still protested against this view of the Supreme Court, asking in the *Münchener Medizinische Wochenschrift* the rhetorical question:

> Why should a doctor who skilfully removes with one cut a malignant tumour, thereby freeing a human being from pain and suffering or even saving the patient's life through his operation, be treated on the same level as a rowdy: *here* is the hostile intention to harm, *there* the will to be useful, to help. ... we have to emphasise firmly the point that we *can never accept our medical and surgical interventions as physical assault and bodily maltreatment in the sense of the Penal Code.*[33]

As these reactions illustrate, the crucial issue for doctors in the Hamburg case was not that one of their colleagues had been found guilty of having operated on a child against parental wishes, but rather the fact that a surgeon had been put legally on a par with the common criminal who has committed an assault. This equation was perceived as a violation of medical honour – not only of the personal honour of the surgeon concerned but also of the collective honour of the medical profession. It was unacceptable to the medical mind to see a doctor in front of a judge, accused as a criminal, for a medically necessary and appropriately performed operation.[34] As Marianne Sinn and Thorsten Noack have recently shown, German doctors continued to fight the legal interpretation of medical interventions as physical injury until well into the 1920s and 1930s.[35]

The issue of consent, in contrast, which had become central for legal interpretation, was regularly downplayed by medical commentators. Thiersch, for example, already claimed in 1894 that patient consent as a precondition for performing an operation had become 'second nature' to medical practitioners, so that exceptions to this rule were 'one of the greatest rarities'.[36] Von Angerer asserted that 'we see every high-handed treatment of a patient as an intrusion into his personal freedom; also for the patient the right to decide over one's own body, over one's own health, must be safeguarded'.[37]

Franz König (1832–1910), professor of surgery in Göttingen, gave assurances that in his clinic 'we regard it ... as illegal to perform any bloody, painful, or dangerous intervention without the specific consent of the patient'.[38]

However, such policy statements were substantially qualified by consideration of the psychological and practical situation of the doctor–patient relationship. Von Angerer, for example, noted that 'often the most educated people are like children in their illness, completely irrational' and wanted what was most wrong for them or had problems in making any decision. In such patients, he suggested, it was more humane not to reveal the seriousness of their disease to them and instead to inform the relatives and to seek support and consent from them.[39] König was even more explicit. It was the patients' task, he demanded, 'to trust and to follow, to silence their own knowledge, their own will' and to subject their bodies to the doctor for repair like a watch to the clockmaker. In emergencies involving children König was prepared to operate against the will of the parents or guardians. Generally, he expected patients to consent to the treatment that the doctor planned to undertake, and he assumed a patient's general tacit consent to all diagnostic and therapeutic measures on the basis of the fact that he or she had voluntarily entered the hospital. If patients refused the treatment that was offered, there was only one solution: they had to leave the hospital.[40]

The medical rhetoric about a patient's right to self-determination thus only thinly disguised a strongly paternalistic, authoritarian attitude and habitus. With the rise of science-based hospital medicine in the nineteenth century, the position of the patient vis-à-vis their doctors had been generally weakened. The ever-widening knowledge gap between medical experts and patients, and the increase in available diagnostic and therapeutic methods of hospital medicine, gave doctors more and more authority in decision-making. The power gradient between doctors and – often lower class – patients seems to have been particularly steep in the university clinics of the Wilhelmine era that surgeons such as König and von Angerer referred to.[41] The militaristic and authoritarian culture of this period pervaded the sick rooms of the large hospitals. While explicit refusal of a medical intervention would normally have been (reluctantly) accepted, patient consent was usually an implicit, tacit or 'silent' matter.[42] The patient was expected to accept as a matter of course whatever treatment or measure the doctor, as the expert, felt necessary or appropriate. Significantly, even the Supreme Court had demanded no more than 'presumptive consent' in its decision of 1894, and it had made no statement at all (yet) about the amount of information that a patient should receive prior to a medical procedure.

However, that a doctor's assumptions about his patient's consent might well be wrong, and that such an error could have unpleasant legal consequences, was demonstrated in 1897 by the high-profile case of the gynaecologist Dr Otto Ihle, who ran a private clinic in Dresden. Examining a woman who was already under narcosis for a curettage of the womb (because of endometritis), Ihle had discovered that she had cystic ovarian tumours. He had therefore immediately opened her abdomen and removed both ovaries and the Fallopian tubes together with the cysts, thus sterilising the patient. When she was subsequently informed about the nature and extent of the operation and the reasons for it, she first appeared grateful. But she later refused to pay the doctor's fee for the operation, and Ihle took legal action against her. The patient, Ms B., now in turn demanded compensation for pain and mutilation. In the subsequent civil trial, the court of first instance, the Dresden District Court, found that Ms B. had to pay a fee, but she successfully appealed. The High Court (*Oberlandesgericht*) of Dresden concluded that Ihle had caused 'intentional and illegal physical injury'. It followed the argumentation of the medical expert in the case, Paul Osterloh, that Ihle could have delayed the operation in order to obtain his patient's consent. Prior to the operation, she had only consented to 'a very small operation'. While the High Court rejected Ms B.'s claim for compensation (because she had 'forgiven' the doctor by initially expressing her gratefulness) as well as Ihle's claim for the fee, it passed the file to the public prosecutor so that he could initiate criminal proceedings according to section 223.[43]

Although the public prosecutor decided not to press criminal charges, Ihle was extremely upset about this whole affair and defended his point of view in public. Referring to the Supreme Court's decision in the Hamburg case, he argued that he had correctly 'presumed' the patient's consent. Moreover, it would have been 'barbaric' to wake up the already nervous and anxious patient from the narcosis and 'frighten and weaken her even more through detailed information about her disease, so that she ultimately won't survive the operation'.[44] Feeling that his professional honour as a doctor had been violated by the Dresden High Court, Ihle petitioned the Saxon *Landtag* to provide at least reimbursement of his legal expenses but this was rejected.[45]

Public responses to the 'Ihle case' or 'Dresden case', as it soon became known, were divided. The Munich medical magazine *Ärztliche Rundschau* defended Ihle's conduct. Also an article in the lay press, in the *Berliner Börsenzeitung*, was sympathetic to his position, stating that the verdict put doctors in a no-win situation: they might be prosecuted for carrying out surgical interventions that turned out to be urgently necessary during narcosis; or they might be punished for failing to take action or for endangering the patient's life through delay of treatment.[46] However, such commentary ignored the fact that this case was not about an extension of an operation, but about a different, large operation, which was based on a new diagnosis and for which no consent had been given. The diagnosis of ovarian cysts did not constitute a medical emergency that would have justified an intervention without consent because of imminent danger. This was pointed out, for instance, by the renowned editor of the *Berliner Klinische Wochenschrift*, the clinician Carl Anton Ewald (1845–1915), who concluded that the verdict of the Dresden High Court was justified.[47] Also Melchior Stenglein (1825–1903), a former Supreme Court judge, accepted the Dresden decision, arguing that in current legal opinion surgery without consent had to be judged as physical injury except in cases of imminent danger.[48] However, Carl Stooss used the opportunity to reiterate his reasons why surgical interventions were not physical injuries in the legal sense, publishing articles on the issue in legal journals as well as in the *Deutsche Medizinische Wochenschrift*, in addition to a substantial monograph on the subject.[49] The *Münchener Medizinische Wochenschrift*, following von Angerer's line of argument, also denied that Ihle was guilty of bodily injury, although it acknowledged that he was not entitled to payment for an operation for which he had not received the patient's consent.[50]

3.5 Patient consent to treatment and to experimentation

Both the Hamburg and the Dresden case had brought the fact to wider public attention that surgery without consent could be classed as physical injury – in criminal law as well as in civil law. In the field of jurisprudence, a flood of publications on the legal nature of medical interventions set in around 1900. This happened partly in the wider context of efforts, from 1902, to achieve a general reform of the German Penal Code (see below). But there were also more specific reasons. In another case, in 1905, the Supreme Court confirmed its interpretation of surgical interventions as factual physical injuries in the sense of section 223;[51] and it dealt with civil law cases in 1907, 1908 and 1911 which illustrated how advisable it was for surgeons to obtain the prior consent of parents or guardians when operating on minors. Without it, they were vulnerable to subsequent claims for compensation, especially when the procedure had not led to the desired result.[52]

It would go beyond the remit of this chapter to discuss the large number of juridical expert opinions that were offered in these and other contexts on the question of the legal status of medical interventions.[53] However, it is important to note that some of the most prominent law professors in Imperial Germany dealt with the problem in considerable detail – in substantial articles or monographs. Two 'camps' of legal opinion on the issue can be identified. One camp, which included Karl von Lilienthal (1853–1927) in Heidelberg, Ludwig von Bar (1836–1913) in Göttingen and Ernst Zitelmann (1852–1923) in Bonn, followed the Supreme Court in regarding the patient's consent as a condition of legal medical interventions. Without consent these interventions had to be seen as physical injury in criminal law or as torts (*unerlaubte Handlungen*) in civil law.[54] The other camp, including Joseph Heimberger (1865–1933) in Strasbourg, Richard Schmidt (1862–1944) in Freiburg and Wilhelm Kahl (1849–1932) in Berlin, argued along the lines of Stooss that it was inappropriate to view surgical treatment as factual bodily harm. For them, consent-seeking was a matter of the doctor's professional ethics and of his personal judgement in the individual case – not a strict legal requirement. A presumption of tacit consent was entirely sufficient.[55] Several variations to these major positions were added in law theses and articles in legal and medical journals. The Strasbourg public prosecutor Werner Rosenberg, for example, suggested applying the German Civil Code's regulations on management without instruction (*Geschäftsführung ohne Auftrag*, sections 677–687) in order to justify treatment without consent in children, the mentally ill or otherwise incapacitated patients in situations of imminent danger. However, his proposal was immediately criticised by other lawyers.[56]

The wide legal interest in the issue of medical interventions and patient consent was also linked with the contemporary publicity around the case of Albert Neisser, the professor of dermatology and venereology at the University of Breslau. Since his case has been examined in detail by Barbara Elkeles, only some major facts of it need to be repeated here. Searching for an immunisation against syphilis, Neisser had injected, in 1892, nine female patients, who were in his clinic for other diseases, with cell-free blood serum from syphilis patients. Some of these test patients were minors, and some were prostitutes. None of them (nor their legal guardians) had been informed about the nature of the injections or been explicitly asked for consent. The experimental injections had been given as if they were part of their treatment. Four test patients – all four of them prostitutes – later developed syphilis. While this made it clear that the serum had no protective value, it also raised the question whether their infection with syphilis had been caused by the experimental injections or whether it had been acquired 'naturally' through their occupation (as Neisser claimed). Initially published as a scientific article, the details of Neisser's experiment were only brought to the attention of a wider public in early 1899 by the liberal newspaper *Münchener Freie Presse*, edited by the historian, social critic and anti-vivisectionist Ludwig Quidde (1858–1941).[57] In this, a distorted account of Neisser's syphilis experiment was given as part of a series of articles on experimentation on 'poor people in hospitals'. Other newspapers quickly took up the 'Neisser case', and it became the subject of lively debates in both houses of the Prussian Parliament and in the medical press. While much of the polemics against Neisser was, as Elkeles has argued, due to anti-Semitic sentiment, there was also genuine concern about the legality of his experiments, especially with regard to the issue of consent. The Royal Disciplinary Court for Civil Servants (*Königlicher Disziplinargerichtshof für nicht-richterliche Beamte*) eventually punished Neisser in December 1900 with a fine of 300 Mark and a reprimand, because he had not obtained the consent of the test persons or their legal representatives and had thus neglected his duties as a physician, director of a clinic and professor.[58]

Alarmed by Neisser's case and other reports about other questionable trials in university hospitals, the Prussian Ministry for Religious, Educational and Medical Affairs commissioned legal expert opinions on experimentation on human subjects. One of the experts was the above-mentioned Göttingen professor of penal law, Ludwig von Bar. In line with his general interpretation of medical interventions as factual physical injuries in the sense of the Penal Code, he regarded scientific experimentation on a person without consent as punishable. In his view, consent could justify experiments that caused minor bodily injury. But such consent could only be given by fully competent adults. Experimentation on minors, even with parental consent, was not permissible. Von Bar also demanded that the volunteering test persons were made aware of the health implications of the proposed trial, although they did not have to be informed about very unlikely, remote risks.[59]

Von Bar's opinion influenced an official instruction to the heads of clinics and hospitals, which was issued by the Prussian Minister for Religious, Educational and Medical Affairs on 29 December 1900 – on the same day that Neisser was disciplined. It ruled that any medical intervention which did not serve diagnostic, therapeutic or immunisation purposes was forbidden, if (1) the person on whom it was carried out was a minor or not fully competent (2) the person concerned had not explicitly given their consent or (3) the person had not been informed in advance about potential detrimental effects. Moreover, the intervention could only be performed by the head of the hospital or clinic or with his special authorisation, and the conditions and specific circumstances of the case had to be recorded in the patient's file.[60]

In other words, information and consent had been made official requirements (in Prussia) for scientific experiments on human subjects. Such experiments could only be performed on competent adults, not on children or adolescents. While it seems that the ministerial instruction had actually little practical consequences for the conduct of scientific research in the hospitals and clinics,[61] it contributed to the general discourse on requirements of consent in medical interventions. Importantly, *therapeutic* interventions had been excluded from the regulations, which meant that the problem of consent to surgery remained unsolved. Also, in the practice of the clinical research of this time the boundaries between therapeutic and non-therapeutic experiments were often blurred.[62] Moreover, when it came to conflicts about patient consent and medical responsibility in operative treatments, surgeons tended to ignore the complex legal literature on the subject or indignantly rejected the position of a legally equal status for surgery and physical injury.[63] Against this background, it is interesting to see what medical authors recommended to their colleagues regarding the question of consent – after the high-profile cases of Waitz, Ihle and Neisser.

3.6 Medical comments and advice on consent and information

Instructive types of literature in this regard are doctors' writings on medical conduct, professional ethics and legal issues relevant to the practice of medicine. Originating from the *savoir faire* (know how) literature of the eighteenth century, this genre of medical writing still enjoyed considerable popularity around 1900.[64] A particularly successful publication of this kind was the booklet *Von Ärzten und Patienten: Lustige und unlustige Plaudereien* (On Doctors and Patients: Amusing and Non-Amusing Conversations), which was first published in 1899 and saw a fifth edition in 1927. Written by the Bremen physician and psychiatrist Friedrich Scholz (1831–1907) in a partly humorous, partly polemical style, it discussed current issues of medicine and gave advice to medical practitioners. In a chapter on the 'limits of medical rights' Scholz reflected on the Hamburg case as well as on the recent debate on experiments on hospital patients. While he criticised potentially harmful scientific experiments such as 'syphilis-injections' and complained about 'the autocracy of some hospital doctors' who regarded dependent patients as 'material' and 'abused them in the service of science', he was supportive of trials of new therapies if cautiously conducted. He inveighed against the Supreme Court's decision of 1894, the legal view of surgical interventions as physical injuries and the power of judges over medical practitioners. However, after the polemics, he admitted that the Hamburg surgeon had infringed the legal rights of the girl's father and that the case might have been a matter for a medical court of honour.[65]

In general terms, Scholz noted that patients entrusted their bodies to their doctor but did not make him 'the master' over them. Although the Bremen physician insisted that the doctor was sovereign in his field and did not need to enter into discussions with the patient, the doctor was not entitled to do anything causing 'great discomfort, pain or danger' without the patient's consent. The doctor, Scholz concluded, had to make sure that he has the patient's permission for the treatment, and he also should not hide potential problems or the seriousness of the illness. Only after such openness would the consent of the patient provide true 'moral support' for the doctor.[66] Although Scholz thus recognised a certain requirement for patient consent, and a limited need for prior information, he remained essentially paternalistic. His information and consent-seeking aimed predominantly at the cooperation of the patient during the treatment. It did not really arise out of respect for patients' autonomous decision-making.

Quite different in tone and conception were the comments by Albert Moll in his handbook of medical ethics, *Ärztliche Ethik*.[67] Moll judged the issue of human experimentation as well as that of consent in surgery on the basis of his general understanding of the doctor–patient relation as a contract relationship. The voluntary consent of the patient to any form of procedure, whether experimental or therapeutic, was therefore essential for him. In fact, the apparent lack of consent (and information) in contemporary clinical research had been the main reason why Moll felt it necessary to publish a book on the 'doctor's duties'. He had collected approximately 600 cases of non-therapeutic human experimentation from the international literature, which he now summarised in his book. In a number of these cases, harmful or dangerous experimental interventions had obviously been made without prior information or consent. Moll left no doubt about his condemnation of such practices:

> ... I have observed with increasing surprise that certain medics, obsessed by a kind of research mania, have ignored the areas of law and morality in a most problematic manner. For them, the freedom of research goes so far that it destroys any consideration for others. The borderline between human and animal is blurred for them. The unfortunate sick person that has entrusted herself to their treatment is shamefully betrayed by them, their trust is betrayed, and the human being degraded to a guinea pig. Some of these cases have happened in clinics whose directors can't talk enough about 'humanitarianism', so that one might almost regard

them as specialists in humanitarianism. There seem to be no national or political borders for this kind of aberration.[68]

Accordingly, he was sceptical about the efficacy of the Prussian instruction. Moll criticised that its text did not make clear enough whether the exemption of interventions for 'diagnostic, therapeutic and immunisation purposes' referred to the individual patient concerned, that is, that these interventions were carried out in the personal interest of this particular individual, or whether the exemption was meant in general terms, covering all procedures of those types. Moreover, he felt that simple recording in the patient's file was no guarantee that objective information had been provided and that the patient had consented to the proposed experiment unambiguously. Moll therefore regarded it as necessary that someone acted as guarantor and that for serious interventions the patient gave his consent in writing. As the Neisser case and other cases had made clear to Moll, in the authoritarian milieu of contemporary hospitals the voluntariness of patients' consent was doubtful. Also, as Moll claimed, most hospital patients were too uneducated to fully comprehend the implications of their consent to an experiment.[69]

With regard to surgical interventions and the Hamburg case, Moll insisted that a patient's right to self-determination had to be respected – from a legal as well as from a moral point of view – even if the refused operation would have been beneficial to the patient's health. Any deviation from this principle would ignore the very basis of the doctor–patient relationship, that is, the contract between the two parties. Moll was aware that his uncompromising position in this matter was closer to the juridical view than to common medical opinion, but he emphasised that a doctor was under no obligation to force a certain treatment onto a patient (except in medical emergencies). If the patient refused a proposed therapy, the doctor was free to break off the treatment, but he was not permitted to act against the patient's will.[70]

The same line had been taken a few years earlier by Moll's colleague Julius Pagel (1851–1912), medical practitioner and historian of medicine in Berlin,[71] in an advice booklet for doctors entering the profession. Like Moll, Pagel quoted the Latin saying *beneficia non obtruduntur* (benefits are not imposed) when discussing non-cooperative patients. The basis for Pagel's conception of the doctor–patient relationship was not, however, respect for the mutual contract or for a patient's right to self-determination. His attitude was openly paternalistic: similar to Scholz, for Pagel the doctor was a 'sovereign' in his medical actions, who simply did not have to put up with unruly patients. While Pagel had taken in the lesson of the Hamburg case, advising his readers that 'in larger surgical operations, especially amputations' the consent of the patient or his relatives had to be sought, he was prepared to forgo information or consent in minor procedures. In such cases even deception was justifiable in the patient's own interest. For example, if an abscess or carbuncle of the back had to be opened, the doctor should divert the patient's attention for a moment, or silently pretend to examine more closely, and quickly make the incisions before the patient fully notices what is being done.[72]

Paternalism was likewise put forward in a 1905 advice book for new medical practitioners, published by the Alsatian physician K. Hundeshagen. In view of the competition from non-licensed healers (the so-called *Kurpfuscher*), who liked to talk with their clients about the nature of their illness, he advised the young doctors to explain to patients, at the bedside, the nature of their disease, its course, and its therapy including the rationale behind the treatment. But at the same time Hundeshagen doubted that many patients were able to truly comprehend and understand such 'education', and he warned against creating misunderstandings by using uncommon terminology or making a gloomy prognosis.[73] Clearly, patient information was valuable for him only in so far as it improved the relationship between doctor and patient and thus contributed to the success of the doctor's practice in a competitive market. Information was not meant to be used to enable patients to make a decision for themselves.

However, the recent court cases on consent, in particular the Hamburg case of 1894, were reflected in further guides on the professional, ethical and legal aspects of medical practice. In 1906, for example, Erich Peiper (1856–1938), professor of internal medicine and paediatrics at Greifswald University, published his student lectures on professional issues. Referring to the recent jurisdiction as well as to the view of 'the experienced Scholz', Peiper stressed that surgical interventions, the preceding chloroform narcosis, and any painful or unpleasant treatments required the patient's prior consent. Patients had a right to be truthfully informed about the implications of a proposed therapy and the chances of success of an operation. If the patient refused the operation, he should be object-ively informed about the consequences, but never be talked into agreeing to it.[74] A handbook on German medical law, published jointly in 1911 by Heinrich Joachim (1860–1933), medical practi-tioner and editor of the *Berliner Aerzte-Correspondenz*, and the Berlin lawyer Alfred Korn, quoted the text of the Supreme Courts's decision of 1894 in full. In their interpretation, this decision meant that con-sent to treatment was generally a requirement in private practice, though exceptions were possible in situations of imminent danger to the patient's health, and in unconscious, mentally ill or legally incompetent patients, where the doctor could treat, under the rules of management without instruction. In view of the ruling of the Supreme Court, it had become customary, as Joachim and Korn noted, to ask the patient or his legal representative for consent to any dangerous operation and to 'even occasionally demand the consent in writing'. The authors however adhered to the conventional opin-ion that the head of a public hospital was generally entitled to any necessary surgical or other procedure, as long as the patient or his representative had not explicitly made this subject to their consent. In those cases where a surgeon had erroneously assumed consent to the operation, and his error was excusable, punishment for negligent (but not intentional) physical injury was possible.[75]

Thus, with the exception of Moll, doctors' comments tended to maintain the premise of medical paternalism and of medical 'sover-eignty' when discussing patient information and consent. Their advice to seek consent and to provide (some) information did not reflect a true recognition of a patient's right to self-determination in matters of their personal health and disease. Rather, such advice was purely practical, aiming at sparing medical practitioners legal troubles or even convic-tions for physical injury.[76] It would appear that the high-profile legal cases on treatment without consent had made an impression on the medical profession, but they had not really changed its basic attitude. The patient was generally regarded as insufficiently competent to evaluate medical matters and was therefore expected to agree to what-ever treatment the doctor regarded as necessary or appropriate. If doc-tors, especially surgeons, asked for explicit, written consent, then this was primarily done to cover themselves in particular circumstances with regard to a potential claim for compensation or charges of bodily harm. This is also evident from the few surviving consent notes or forms dating from the early twentieth century: they are either very short or include a clause that gives consent to any further interventions that may become necessary during an operation.[77]

3.7 Discussions and efforts within the penal law reform

As has been shown above, doctors adapted to the legal situation after the Supreme Court's decision of 1894 by giving practical advice on how to safely handle the matter of patient information and consent. However, considerable resentment continued to exist about the legal interpretation of medical interventions as factual physical injuries. In July 1902, the German Penal Law reform began with the formation of a scientific committee of seven legal experts (from both the 'classical' and the 'modern' school of penal law) and one psychiatrist (Gustav Aschaffenburg) by State Secretary Arnold Nieberding of the Reich Justice Office.[78] The subsequent public discussion about the legal reform provided the medical profession with opportunities to argue for a change in the Penal Code that would exempt medical procedures without consent from being punishable under section 223. Particularly active in this regard was the medical chamber of Brandenburg–Berlin, which formed a subcommittee for this task in the same year, 1902. Its proposal, published in 1905, suggested the insertion of two clauses into the Penal Code, which would have exempted from punishment any medical intervention of a licensed doctor (*approbierter Arzt*) that was performed *lege artis* and that was (1) not in conscious conflict with the free decision of the patient or his legal representative, or (2) necessary to save the patient from a present danger to his body or life.[79] In essence, this proposal would have left only a medical intervention against the explicit will of the patient punishable. Presumed tacit consent would have fully protected the doctor against legal charges, and active consent-seeking and provision of information for the patient would not have been requirements. Such protection would have been guaranteed to licensed doctors only, whereas non-licensed healers, who had freedom to practice (*Kurierfreiheit*) under the German Trade Ordinance of 1869/71, would have been unprotected.

The medical chambers of Pomerania, Schleswig-Holstein, Hamburg and Hesse-Nassau agreed with the Brandenburg–Berlin proposal. The chambers of Westphalia, the Rhine Province and Hanover suggested, in 1906, to insert a new clause in the Penal Code that would only make high-handed, deliberately non-consensual medical treatments (*eigenmächtige ärztliche Behandlungen*) punishable under the section on wrongful deprivation of personal liberty (section 239).[80] However, some legal experts severely criticised these medical proposals. According to Bonn professor Ernst Zitelmann, for example, the notion of a present danger to the body was far too vague and would 'surrender' every human being ('since nobody is completely healthy') to any licensed doctor. He also disapproved of the implicit exclusion of non-licensed healers and wanted to see consent and management without instruction explicitly mentioned in a reformed Penal Code as the reasons for the legality of medical interventions.[81] Wilhelm Kahl, who was a key member of Nieberding's scientific committee, revived Binding's notion of a professional right (*Berufsrecht*) of licensed doctors to perform surgical as well as other treatments. This professional right, he argued, was recognised by the state, as the licensed doctor practised in the public interest of health care. Therefore, medical interventions which had been carried out *lege artis* were not punishable, and it was thus superfluous to include a special clause on this in the Penal Code.[82]

In fact, when the first penal law reform commission (formed by Nieberding in 1906 and consisting of five judges and ministerial civil servants) published their Preliminary Draft Code (*Vorentwurf*) in April 1909, none of the medical proposals had been integrated. The commission saw no need to change section 223 in this respect, and they rejected the proposal to include a specific regulation on high-handed medical treatment, because the general section on coercion (section 240) was thought to provide sufficient protection for patients.[83] However, the organised medical profession did not give up. The topic of consent and physical injury was put on the agenda of the thirty-eighth assembly of the German Association of Doctors' Societies, held in Stuttgart in June 1911. Salomon Alexander (1852–1928), secretary of the medical chamber of Brandenburg–Berlin and member of its relevant subcommittee, reported about the unsuccessful efforts to influence the penal law reform and attacked the persistent legal notion of surgical interventions as factual physical injuries – a notion which he believed to stand in sharp contrast to doctors', as well as the general public's, view of the 'dignity and social utility' of the medical profession. Alexander however supported a clause on help in emergencies (*Nothilfe*) in the Preliminary Draft Code that exempted from punishment interventions – even against the person's will – which were undertaken to save the person from a considerable danger. A second speaker, the Essen physician Friedrich Heinsberg, echoed Alexander's views, complaining of a violation of the dignity and reputation of the profession. Heinsberg warned that German doctors might face 'American circumstances' with regard to legal charges against doctors, and he accused colleagues using consent forms of undermining the relationship of mutual trust between doctor and patient. Following a brief discussion, the assembly agreed on a resolution according to which appropriately conducted medical interventions must not be judged as physical injuries. The current legal insecurity, it was claimed, hindered medical practice and thus harmed public health.[84]

Commenting on the *Vorentwurf*, Joseph Heimberger, who in the meantime had become professor of criminal law in Bonn, and Ludwig Ebermayer, as a Supreme Court lawyer, both suggested the incorporation into the new German Penal Code of a specific clause on high-handed, non-consensual medical interventions, as Austria had done in section 325 of its draft penal code. This clause would have punished interventions against the will of the patient or his legal representative, but excluded from punishment interventions that were made to save the patient from an immediate danger to his life. Such a clause had also been proposed in a 'counter draft code' (*Gegenentwurf*), issued by a group of prominent professors of law, including Kahl, von Lilienthal, von Liszt and James Goldschmidt.[85] Thus the 1906 proposals of the medical chambers of Westphalia, the Rhine Province and Hanover had found at least some indirect support from important legal experts.

In October 1911, a newly formed grand penal law reform commission (*große Strafrechtskommision*), consisting of 15 jurists appointed by the Reich government, conducted the first reading of the proposed clause on help in emergencies. It had been put on the agenda by Ebermayer, who had become a member of the grand commission. However, the pleas of the medical representatives, including Salomon Alexander, were less effective than those of the

Göttingen professor of law, Robert von Hippel. The commission followed von Hippel's opinion that, even in emergencies, medical interventions must not be performed against the declared will of the patient. A commission meeting in July 1912 did not move in favour of the medical interests either. Ebermayer proposed to explicitly exclude medical interventions from the regulations on physical injury (section 223) and to introduce a new specific clause on punishment of high-handed medical treatments, thus reflecting to a large extent the wishes of the various medical chambers. But, again following a plea by von Hippel, such a change to section 223 found no majority; and in the second reading, in March 1913, both Ebermayer's proposal of a clause on high-handed medical treatments and a further proposal on medical help in emergencies were rejected.[86]

Thus, nothing changed about the penal law status of medical interventions. Medical treatment without consent could still be punished as physical injury. Even before the outbreak of the First World War put a temporary stop to the work of the penal law reform commission, the efforts of the medical profession, which had partly been mediated by Ebermayer, had failed.

However, while the requirement of consent thus remained firmly in place, the Supreme Court displayed a more conciliatory attitude to the question of patient information. In a landmark decision on 1 March 1912, the highest court had, for the first time, dealt with the question how extensive and detailed patient information before surgery should be. In the relevant civil law case an ear operation (carried out with the patient's consent) had gone wrong, resulting in permanent deafness of that ear. The patient's claim for compensation rested in part on the accusation that the doctor had informed him insufficiently about the risks of the operation, and the High Court (*Kammergericht*) of Berlin had accordingly convicted the doctor. The Supreme Court, however, suspended the verdict, arguing that there was no obligation for the surgeon to inform the patient about remote risks.[87] Contrary to its earlier insistence on a patient's right to self-determination its decision reflected the continuing medical paternalism:

An obligation of the doctor to draw the patient's attention to all disadvantageous consequences that might possibly follow from the recommended operation cannot be recognised. The assumption of such an obligation can neither be derived from the practice of dutiful and careful medical practitioners, nor from internal reasons. Providing comprehensive information to the patient about all possible disadvantageous consequences of the operation would often be actually wrong, if the patient is deterred by this from undergoing the operation, although it is, despite its risks, necessary or at least advisable, or if the patient, by imagining the dangers of the operation, is put in a state of fear and agitation and a good course of the operation and of the cure is thus endangered.[88]

3.8 Conclusions

On the eve of the First World War, then, the issue of patient information and consent was in an unclear state. On the one hand, the legal interpretation of medical interventions as factual physical injuries and the court decisions supporting this view had forced doctors to assure themselves of their patients' consent. Moreover, scandals around dangerous human experiments in hospitals had led to the Prussian regulation of 1900 requiring patients' information and consent in non-therapeutic trials. On the other hand, requirements regarding the extent of patient information remained minimal, and evidence of tacit consent to treatment was still sufficient from a legal point of view. In such a situation, doctors were able to adhere to their traditionally paternalistic perspective and habitus, regarding themselves as 'sovereigns' in medical decision-making. Serious, especially surgical interventions against the explicitly declared will of the patient had been recognised by many doctors as something that was legally problematic. But assurances that they respected a patient's right to self-determination over his own body were largely rhetorical. The medical profession had mostly reacted to legal pressures by defending its position and honour and by making only a few practical concessions. There was little change in attitude. The agenda of strengthening the patient's role in matters of health care had largely been driven by lawyers, not by doctors.

During the First World War, a brief discussion emerged to what extent soldiers had to tolerate surgical interventions that were necessary to restore their fitness for military service. Relevant to this was a verdict of the Reich Military Court in 1915, which endorsed such a duty of soldiers for medically necessary, minor operations as part of their general duty of obedience. It preserved, however, a requirement of their consent for 'major' interventions.[89]

4
Duties and Habitus of a Doctor: The Literature on Medical Ethics

4.1 Introduction

Writing about the doctor's duties has a long tradition in Western medicine, reaching back in principle to the Hippocratic Corpus and, within it, especially to the Hippocratic Oath (*c.* 400 BC).[1] In many periods, such writing has been employed not only to provide ethical guidance for medical practitioners, but also to demarcate their conduct from that of other (often rival) healers and to introduce young doctors to forms of 'prudent' behaviour vis-à-vis patients, medical colleagues and other practitioners, and the wider public.[2] The art of acting prudently, in order to build up a successful medical practice, was especially cultivated in the *savoir faire* (know how) literature of the Enlightenment, from Friedrich Hoffmann's *Medicus Politicus* (Latin 1738, German 1752), via John Gregory's *Lectures on the Duties and Qualifications of a Physician* (1772, German 1778) and Wilhelm Gottfried Ploucquet's *Der Arzt* (1797), to Thomas Percival's *Medical Ethics* (1803) and Christoph Wilhelm Hufeland's essay 'Die Verhältnisse des Arztes' (1806).[3] As Iris Ritzmann has recently argued, this type of advice literature was meant to help young practitioners cope with uncertainty and insecurity, in a time when the therapies of academic medicine still had little more to offer than the various empirical treatments of other healers.[4]

The tradition of writing about medical deontology, that is a doctor's duties, continued throughout the nineteenth century and beyond, but then needs to be read against the background of the rise of scientific medicine and medical professionalisation, and the specific problems related to these developments in different national contexts.[5] Having charted relevant institutional and legal dimensions of medical professional ethics in Imperial Germany in Chapters 1 to 3, I will discuss in this chapter the ethical views that were expressed in contemporary writings about the doctor's duties. In my analysis of this literature I draw upon Pierre Bourdieu's concept of 'habitus' as a specific pattern of perception, thought and behaviour of a social group.[6] It will be shown that the writings on medical deontology created and propagated an ideal habitus for doctors. The implications of the ethical attitudes as well as of the ideal habitus will be illustrated with regard to some common issues in the medical practice of the time, including truth-telling at the bedside, care for the dying and abortion. Equally as revealing are the views of medical practitioners on the role of science in medicine.

4.2 The conservative critique of K. F. H. Marx

As discussed in Chapter 1, the Trade Ordinance of 1869/71, which put medical practice on a par with other trades (*Gewerbe*), caused considerable resentment among conservatively minded doctors, whose self-perception as academics and members of the middle class conflicted with this new status. One of these critics was Karl Friedrich Heinrich Marx (1796–1877), professor of medicine at the University of Göttingen. A prolific though somewhat idiosyncratic writer, he published several collections of aphorisms and monographs in response to this issue, including, in 1874, a critique of the 'habits and directions of present-day physicians' and, in 1876, a 'medical catechism', that is an outline of the doctor's professional obligations.[7]

Although Marx acknowledged that the Trade Ordinance followed the logic of modern economics and that the new method of charging for specific services (rather than waiting for an annual honorarium) might suit financially pressed practitioners, he deplored the fact that the regulations had reduced doctors to 'salaried employees'. Medicine, he claimed, was as little a trade as theology, and a doctor's practice could not be seen as a business with a changing market value. Instead, the doctor was supposed to be a careful and conscientious preventer and healer of illnesses, a consoler and protector of the sick.[8] His ideal doctor was endowed with scientific as well as ethical perceptiveness, with an independent, critical mind, and a 'moral aristocracy of character'.[9] For Marx, the natural sciences, such as physics, chemistry, botany and zoology, were overrated in medical education, as were pathological anatomy and the new methods of physical diagnosis, bacteriology and experimental physiology. The last method was also criticised by him on moral grounds because of the animal suffering involved in vivisections and their supposed hardening effect on experimenters and students. What Marx advocated instead was the role of the doctor as a 'healing artist' (*Heilkünstler*), the command of semiotics (i.e., of the traditional doctrine of the signs of diseases), early experience of medical students at the sickbed and the study of general pathology, therapeutics and the history of medicine.[10]

On the basis of this anti-modernist critique of medicine, Marx elaborated upon the habitus and duties of the doctor, or as he put it, 'how the healing artist should be and act in order to do justice to the high task of his profession'.[11] Ideally, Marx thought, medicine should be practised by men of private means, as they could help the sick out of a pure sense of compassion and moral duty, without a need to consider their fees. Moreover, their independent status would allow them to exercise objective judgement and to remedy abuses.[12] Medicine, he stated, was 'a part of ethics'.[13] In line with this understanding of medical practice Marx denounced the eighteenth-century *savoir faire* as 'paltry arts', which in his time were seen as 'the knowledge of how not to do it'.[14]

As far as his relationship with his patients was concerned, Marx's ideal doctor was considerate and steady, and courageous in treating contagious diseases. The doctor had to be absolutely discreet about the patient's condition and personal circumstances – Marx called the doctor's professional secrecy a 'natural law' (the fact that it was regulated in section 300 of the German Penal Code of 1871 was not mentioned by him). Even if the doctor was unjustly accused of malpractice, he had to keep silent.[15] Marx's position on the question of truth-telling was typical of his time. A dire prognosis had to be withheld from the patient in the latter's own interest: telling the truth in such a situation might even lead to a patient's suicide.[16] Apart from this, there was the risk that the doctor may have erred about the seriousness of the patient's condition.[17]

In terminally ill patients, the doctor should ease their last days with loving care and palliative treatment, providing them with 'euthanasia', that is a 'good', peaceful death in the traditional sense of the term. Marx referred here to a separate essay of his specifically on this topic, which he had presented at his inaugural lecture as extraordinary professor at the University of Göttingen in 1826. As he had made clear in his lecture, such care for the dying did *not* involve actively shortening their lives (in the sense of active euthanasia), but could include restriction of treatment to basic measures and the giving of opium or other narcotics for the relief of pain and discomfort. The decision over life and death lay with a 'higher power', and it would be 'horrible' if the physician, whose task was to preserve live, believed himself entitled to hasten the patient's end.[18] This understanding of 'euthanasia' as palliative care was entirely in line with that of other medical writers of the early and mid-nineteenth century who discussed this topic in detail.[19] Such a conception of a 'medical euthanasia' existed partly in contrast to, and in deliberate demarcation from, traditional lay practices, which could include active measures for hastening death, such as abruptly pulling a dying person's pillow from underneath her or suffocation by pressing a cushion over her face.[20]

To protect his honour, Marx's ideal doctor had to resist amoral requests from patients, such as an aphrodisiac or an abortifacient. A request for the latter should not only be firmly rejected, but accompanied by the threat to inform the authorities should one subsequently hear of a miscarriage. Losing good relations, recommendations and income through such behaviour was less important than a loss of self-esteem and professional honour.[21] In general terms, Marx placed the doctor's professional honour (*Standesehre*) even higher than his personal honour. Collegiality was of great importance to Marx: any public criticism of the treatments of another doctor was forbidden, and 'small deficits' had to be excused or overlooked. Membership in a medical society appeared to him a good way of cultivating mutual respect and a sense of community among doctors.[22]

In sum, Marx built his notions of medical professional ethics on a deep sense of moral responsibility and personal as well as professional honour. His ideal doctor's habitus was that of the serious medical scholar and humane helper of the sick. Other than in the *savoir faire* literature, personal success through smooth and prudent behaviour was unimportant to him. In his conception of medical learning he adhered to the traditional art of individualised diagnosis and treatment and remained sceptical about the role of the new experimental, scientific approaches to medicine. His biographers' judgements about his influence were divided. While Heinrich Rohlfs, a student of Marx, eulogised him as a 'founder of ethical medicine', the Göttingen professor of pharmacology, Theodor Husemann (1833–1901), claimed that 'demonstrably, Marx's "ethical striving" had no significant or permanent practical success'.[23] Clearly, Marx's ethical reasoning and medical deontology were those of an elite physician – he had become ordinary professor in 1831 and *Hofrat* in 1840. In contrast, from the 1880s much of the advice literature on doctors' duties and conduct was published by less-known medical practitioners.

This broadening of authorship correlated with the development of certain issues for the medical profession, which had been aggravated by the Trade Ordinance: more open competition with lay healers; increasing numbers of medical students and 'overcrowding' of the medical profession; and a debate about the admission to the study of medicine of school-leavers from *Realgymnasien*, that is from high schools with a focus on natural sciences, technology and modern languages, in addition to the graduates from the traditional humanistic *Gymnasien* with their focus on the classical languages and historical studies.[24] In this situation, the literature on a doctor's role and duties addressed medical practitioners (especially those new to the profession) and actual or prospective medical students as well as the general public.

4.3 Medical practitioners' views in the 1880s

Several of the above-mentioned issues, for example, were reflected in a booklet by a Silesian medical practitioner, Dr Heinrich Schmidt. Published *c.* 1884, it complained about the consequences of the Trade Ordinance, especially about the competition from naturopaths and homoeopathic lay healers, but also about the competition between licensed medical practitioners themselves. According to Schmidt, honourable conduct of doctors, which had hitherto earned them respect and trust in the population and built a 'protective wall' against quacks (*Kurpfuscher*), was no longer sufficient.[25] The geographical concentration of doctors, he claimed, did not follow the density but the affluence of the population. While a major city such as Leipzig had then approximately 1 doctor per 800 inhabitants, a poor industrial town such as Chemnitz had 1 per 3,500. Compulsory health insurance for workers, introduced by law in 1883, had not yet produced much effect: Schmidt observed that large parts of the population simply could not afford to consult a doctor, and he therefore proposed that fees be fixed at a low level. The number of practising doctors should be regulated according to the regional population density, and all practitioners should have a uniform medical education at a university for five years, followed by one year of practical training at a hospital or in a medical practice.[26]

In contrast to Marx in the previous decade, Schmidt expressed strong support for the new scientific approaches to medicine, especially bacteriology and experimental physiology, which he expected to lead to substantial therapeutic advances. Writing at the height of the first German debate about animal experimentation (in 1885, a ministerial decree regulating vivisection was issued in Prussia), he saw the anti-vivisectionists' campaign as 'wrong sentimentality' and as a sign of 'utter ignorance'.[27]

While Schmidt's arguments and proposals were largely of a political nature, other writers of this time adhered more closely to the tradition of advice literature for young medical men,[28] including also advice for patients. A booklet entitled *Arzt und Patient. Winke für Beide* (Doctor and Patient. Hints for Both) was published anonymously in 1884 by the Viennese surgeon Robert Gersuny (1844–1924) with Ferdinand Enke Verlag in Stuttgart, a medical publishing house, which had also brought out some of Marx's works on medical ethics. A second edition with the same publisher, but now openly under Gersuny's name, came out in 1896, and a further edition appeared in 1904. Gersuny depicted the ideal habitus of a paternalistic doctor, who had to cope good-naturedly with suspicious, non-cooperative, irrational or overanxious patients. An important implication of this paternalism was that the doctor was advised to be very restrictive in giving information to patients with serious illnesses. Certain diagnoses, such as cancer or tuberculosis, were unmentionable because they would deprive patients of hope. Caution had also to be exercised in speaking with the relatives, as they might pass on the information to the patient or might become unable to continue as carers because of their sense of hopelessness. Discretion was further demanded in diseases which might have been caused by another member of the

family, for example through venereal infection or through careless-
ness. According to Gersuny, it was best if the doctor led patient and
relatives 'silently past the danger'.[29]

The ideal habitus of the patient, complementing the paternalistic
habitus of the ideal doctor, was that of complete trust and of a seeker
for help.[30] To maintain this trust, strict confidentiality was essential.
Requests for information by acquaintances of the patient should be
met with the remark 'I don't talk about you to others either', thus creating the image of the doctor who never reveals anything about his
patients to others.[31] Similar to Marx, Gersuny also advised to reject
firmly any morally problematic requests, for example for a medical
certificate to protect a young man against the draft for military service or for covering up a pregnancy resulting from adultery. Apart
from morality demanding such firmness, the doctor was thought to
be well advised to avoid any appearance of corruptness.[32]

Similar in scope, but somewhat different in style and focus, was
an advice booklet for incoming medical practitioners published
in 1886 by Dr Wilhelm Mensinga of Flensburg under the pseudo-
nym 'Dr. med. C. Hasse'. Weaving autobiographical elements into
his text, Mensinga discussed in particular detail the educational
requirements for future medical doctors and the ethical conduct of
practitioners whose work involved obstetrics and women's diseases.
In the context of the controversial issue of opening up medicine to
students from the *Realgymnasien* (which was resolved in 1900 by an
Imperial decree which made them eligible),[33] he highlighted the
advantages to medical studies of mathematics, the natural sciences,
modern languages and of drawing skills. A classical education, he
claimed, had no greater influence on the student's moral and religious development than other forms of education. What the critics of the *Realgymnasien* really feared, according to Mensinga, was
a loss of reputation of the medical faculty compared to the other
faculties (law, theology and philology) if it admitted students without the prestigious background in classical studies.[34] At this time,
Mensinga's position on this question was still that of a minority
within the medical profession. However, his advice to study medicine for idealistic rather than economic reasons was entirely in line
with the profession's efforts to prevent overcrowding by trying to
dissuade school-leavers from choosing medicine.[35] In fact, the number of medical students at German universities had risen from 3,333

in 1875/76 to 7,680 in 1885/86. Despite warnings against studying medicine, it reached a peak at 8,558 in 1889/90.[36]

Thoroughly paternalistic, Mensinga described the habitus of the doctor as that of an idealist, humane and compassionate helper and protector, especially of female patients. Touching upon the contemporary debate on admitting women to medical studies,[37] he suggested that the call for female doctors was superfluous if medical men followed this ideal. Given that abortion was illegal under the German Penal Code of 1871, it is not surprising that Mensinga only vaguely hinted at it in his book, speaking of the doctor's role as a 'friend and mediator' for 'fallen' women. Quite openly, in contrast, he argued for 'facultative' (voluntary) sterilisation of women for whom a further pregnancy would pose serious health risks, for example after a Caesarean section.[38] Discretion was especially important to him with regard to women's diseases and as the basis of patient trust. Hence he held that breaches of medical confidentiality were rightly punishable under the law.[39] His ideal doctor exercised deep empathy with women, and yet he was also a valued participant in the rituals of male culture: proudly Mensinga remembered his role as a medic in attendance at student duels (*Paukarzt*) and his own duelling in his fraternity.[40]

Besides monographs such as those discussed above, published academic lectures provided an opportunity to prepare students for their future role as medical practitioners and to reach a wider audience as well. For example, in 1887 a lecture for clinical students on the tasks of the medical profession, given by the Munich professor and director of the Medical Clinic, Hugo von Ziemssen (1829–1902), was published as part of his series *Klinische Vorträge*.[41] Overcrowding of the medical profession and the debate on the admission of school-leavers from the *Realgymnasien* were also the immediate contexts for this lecture. In the latter issue von Ziemssen took the opposite position to that of Mensinga. In his opinion, the classical humanistic *Gymnasium* was the best form of education at the time, as it gave students an 'idealistic drive' and a training in thought and judgement which would later elevate their medical practice beyond a craftsman's occupation. Von Ziemssen belonged to those who feared overcrowding of the medical faculties, the creation of a 'medical proletariat' in the cities, and a loss of reputation for medical doctors compared to the other academic professions.[42] His adherence to the view that a classical education was relevant to a young man's later behaviour as a doctor reflected a belief which was widely held among academics in late nineteenth-century Germany: that the humanities (*Geisteswissenschaften*) gave a basis for social values that could not be provided by the natural sciences.[43]

With regard to the ethical conduct of doctors, von Ziemssen's views did not differ significantly from those of less-known medical practitioners such as Schmidt, Gersuny or Mensinga. Particularly strong was his call for a 'serious' habitus of the doctor: 'Show in your appearance the *seriousness*, which the profession demands, in all things', he exhorted medical students and future doctors, 'in [your] posture, clothes, speech, and especially in your dealings with the sick and their relatives'.[44] Coupled with this seriousness was discretion, as well as providing emotional support in cases of incurable illness. Hope had to be maintained in the patients and their families, without compromising honesty.[45] Continuing education of medical practitioners, to keep up with the progress of modern scientific medicine (e.g., in bacteriological diagnostics and antiseptic wound treatment), was likewise portrayed as a duty, but also as a necessity in order to maintain one's reputation among the public and to stay competitive.[46] Consultations with other doctors in difficult cases were advocated by von Ziemssen as a means of sharing responsibility and reassuring the patient. However, the doctor was not supposed to strive for popularity among the lay public – he simply had to accept that homoeopaths, naturopaths and magneto-therapists might be consulted as well by his patients.[47] Regarding collegiality, von Ziemssen praised the positive effects of medical societies on professional consciousness (*Standesbewußtsein*) and solidarity, strongly encouraging his students to join such a society at their future place of practice. In particular, he observed that strict organisation had helped doctors to be taken seriously by the government authorities and by the courts as experts in matters of health and disease.[48]

There was thus, in the medical–deontological literature of the 1880s, a considerable uniformity of views. While opinions differed on the question of the most suitable school education for future medical students and doctors, the responses to the ethical challenges of the time were very similar. Fearful of overcrowding of the profession and of competition from lay healers, authors conjured up the ideal of the doctor as a paternalistic helper of the sick and their families, as a practitioner who was committed to the new scientific medicine and as a moral, truthful, serious and discreet man.[49]

4.4 Doctors and their 'struggle for existence': Medical deontology and teaching medical ethics in the 1890s

In the course of the 1890s, the competitive climate for medical practitioners became even more accentuated. Between 1887 and 1898, the average number of doctors in German cities rose from 72.6 to 89.4 per 100,000 inhabitants.[50] Applying the famous Darwinian notion, Hugo von Ziemssen had already spoken in 1887 of the doctors' 'struggle for existence' (*Kampf ums Dasein*), when he justified their need for appropriate pay and defended them against reproaches that they had become less 'humane'.[51] Likewise, the Tübingen professor of medicine and medical history, Hermann Vierordt (1853–1943), referred in a public lecture in 1893 to 'the struggle for existence, about which one has to hear and read so much, and in which the medical profession has now also become involved'.[52] By 1896, *Kampf ums Dasein* had become a subsection in an advice book for new doctors, published by the Berlin medical practitioner Jacob Wolff. Referring to the situation in major cities, Wolff observed, 'Here, the *struggle for existence*...intensifies! And as everywhere in life and nature – the stronger one drives out the weaker. The weaker one perishes or tries to develop an activity appropriate to his powers under different or more favourable conditions.'[53]

One consequence of the perceived increase in competitiveness among medical practitioners was, as I have shown in Chapter 1, the creation of medical courts of honour in order to enforce professional discipline. How did writers on medical deontology, such as Wolff, respond to the increasingly competitive climate? Generally, the literature of this type continued to advise on how do deal effectively and honourably with patients, but more room was now given to hints on how to gain a good reputation in public and on how to attract and keep patients without violating the rules of collegiality. This kind of advice was very prominent in Wolff's book for young practitioners, for example: never to admit or show insecurity vis-à-vis patients; to provide them with a prescription even if dietary or hygienic measures would have been sufficient; to make oneself available for home visits during weekends and holidays; to keep in mind that women were most important for spreading a practitioner's name and for recommending him as a family doctor, so that obstetrics was the foundation of a successful medical practice.[54] Moreover, the young medical practitioner was advised to aim for a fixed panel-doctor position as a basis for his income, at least as long as he had not yet built up a sufficiently large private practice. Locum practice was another opportunity to make oneself known among patients. While poaching patients was forbidden, 'attrition' from the clientele of older and established colleagues was deemed natural and acceptable.[55] Unsurprisingly in view of the competitive situation for medical practitioners, Wolff gave much space to advice on how to obtain tactfully but effectively one's honorarium. In general, his recommendation was to act as a 'gentleman' in calculating the fees and to adjust them to the patient's financial circumstances. However, giving discounts to certain groups of patients or underbidding in applying for a new panel-doctor post were regarded as businesslike behaviour that was unworthy of the medical profession.[56]

In common ethical problems, such as abortion and dealing with incurable patients, Wolff's views were similar to that of the practitioner-authors of the 1880s discussed above, but more explicit. Requests for an abortion should calmly but firmly be rejected by pointing out that it was punishable with imprisonment and by describing the moral and physical risks of the procedure. Unmarried women should be consoled, to married women with many children hints could be given about preventing future pregnancies after the present one was completed, that is about the possibility of sterilisation or other contraceptive methods. If the doctor was called to a woman who was miscarrying after an abortion attempt, he should help her without asking any questions about the 'why' or 'where'.[57] Such discretion, as recommended here by Wolff, was in line with doctors' views about medical confidentiality[58] and was reflected in the relatively low numbers of criminal convictions due to performance of an abortion. Although abortion was punishable as a crime with imprisonment or penal servitude of up to five years under the Penal Code of 1871, and doctors had to report preventable crimes, an estimated number of 300,000 to 500,000 illegal abortions were carried out in late nineteenth-century Germany every year, but on average less than 1,000 cases per year ended with a criminal conviction (between 1882 and 1912).[59]

Regarding truth-telling, Wolff was clear that in malignant diseases, for example carcinoma of the stomach, the diagnosis had to be withheld from the patient. In order to preserve the patient's hope some 'pious deception' (*pia fraus*) was permissible. The relatives, though, could be informed about the true nature of the disease and its dire prognosis.[60] Generally, however, doctors had to observe strict confidentiality, not only as a moral duty towards their patients but also in order to maintain their authority and dignity. Talkativeness could even endanger their professional existence as medical practitioners.[61]

In many respects, Wolff's detailed practical advice, given against a background of strong intra-professional competition, had again brought the deontological literature closer to the eighteenth-century writing on *savoir faire*, which had been despised by Marx in the 1870s. In 1897, just a year after Wolff's book, Julius Pagel's guidebook for incoming medical practitioners was published. Pagel placed his deontology firmly in the tradition of writers such as Friedrich Hoffmann. This was not simply a result of Pagel's historical interests. Dwelling for several pages on the contents of Hoffmann's *Medicus Politicus*, Pagel praised this work as paradigmatic and as a 'literary milestone', even though some aspects, such as conduct during consultations, had been treated rather too briefly in his view.[62] Clearly, there was now a good market for advice books for young doctors. As Pagel observed, Wolff's book, published by the Enke Verlag in Stuttgart, had sold out within months, and the same publisher had just brought out the second edition of Gersuny's booklet.[63] Only a couple of years after Pagel's guide book came out, the Berlin publishing house of August Hirschwald brought out yet another booklet of this type, a *Viaticum*, consisting of the collected 'experiences and advice of an old doctor', Carl von Mettenheimer, edited posthumously by his son Heinrich. Written in the tradition of the *savoir faire*, it even included a section entitled 'Mundus vult decipi' (the world wants to be deceived).[64] In the same year, 1899, a German translation of the experiences of an 'old' Italian doctor, J. B. Ughetti of Catania, was published. This book saw at least three editions.[65] And likewise in 1899, Friedrich Scholz's popular 'conversations' on doctors and patients (see Chapter 3) were published for the first time – they went through five editions, the last one in 1927.[66]

Pagel's *Medicinische Deontologie*, subtitled – similar to one of Marx's books – 'a small catechism for future practitioners', was based on a series of articles which he had published during the year 1896 in the *Allgemeine Medicinische Central-Zeitung*. Pagel had been commissioned by the journal's editor to write these articles, but the main cause was a discussion in Berlin about officially introducing the teaching of medical ethics at the university. A need for such teaching had been identified in a meeting of the doctors' chamber for Brandenburg–Berlin in February of that year. Actually, aspects of this subject were already being taught by the medically qualified Berlin professor of philosophy, Max Dessoir (1867–1947), in a course on applied ethics and the higher professions,[67] and by Pagel himself in his lectures for first-year medical students on 'encyclopaedia and methodology of medicine'. Although Pagel's introductory lectures were popular and also attracted students from other faculties,[68] little or no notice had been taken of them by the local representatives of the medical profession. His lectures advocated a materialistic world view and expressed a strong commitment to the natural sciences as the basis of medicine.[69] What the members of the doctors' chamber desired, however, was very practical guidance for qualifying or recently qualified doctors, including topics such as appropriate conduct among colleagues, towards other health personnel, apothecaries, druggists and manufacturers, and towards patients, especially within the context of the health insurance system. While Pagel held the view that the university was not the right place for instruction in medical formalities and tricks of the trade, he had to defend his standing as a medical practitioner (with, as he stressed, 20 years experience) and as a university lecturer. Thus he provided, with his article series and subsequent book, the kind of practical guidance that the doctors' chamber had called for.[70]

While Wolff's book had focused on the young doctor who wants to establish a practice in a major city, Pagel's *Medicinische Deontologie* covered medical practice in a town as well as in rural regions, the role of the poor law physician, of the panel doctor and of the health officer. Consultations, medical societies and professional discipline were discussed as was the doctor's conduct in dealing with apothecaries, druggists, chemists, medical assistant personnel, midwives and with non-licensed healers. Advice on fee-taking and medical bookkeeping was also included. For Pagel, a doctor had to regulate

his life and behaviour in every regard according to a 'threefold honour': he had to maintain his individual honour as a man, his dignity
as a member of an academic profession and his 'specifically medical
class consciousness'. Pagel compared medical honour with that of
the military officer and suggested that the honour of doctors placed
even greater demands of responsibility on them, because doctors had
to determine their conduct without detailed orders from a higher
command.[71] Given the very high social status that officers enjoyed
in Wilhelmian Germany, this was certainly strong rhetoric.

The habitus of Pagel's ideal doctor included not only a benevo-
lent disposition but also a certain social distance from his patients
(especially if they came from the lower social classes), discretion,
industriousness, observance of hygiene, a sober lifestyle and sexual
restraint.[72] In his relationship with his patients, the doctor was supposed to take the role of a 'sovereign' and was always to display full
confidence in his diagnosis and treatment. Pagel's paternalism was
likewise reflected in his advice to be lenient with patients and to
regard them as 'also psychologically affected, often relapsing sinners
against hygiene'.[73]

Despite Pagel's efforts, and the considerable amount of advice literature for young doctors that had been published in the 1880s and
1890s,[74] the call for teaching medical ethics at the universities was
made again in 1898, at the Congress for Internal Medicine held in
Wiesbaden. Oswald Ziemssen, a local medical practitioner and cousin
of the Munich clinician Hugo von Ziemssen, proposed to integrate
training in medical ethics into the clinical curriculum. Lawyers, busi-
nessmen and military officers, he claimed, all instructed their new
blood in their specific ethical views, while current problems in the
medical profession stemmed from a decline of 'ethical sentiment'.[75]
This suggestion was firmly rejected in a reply by Rudolf von Jaksch
(1855–1947), professor of internal medicine at the German University
in Prague. He denied that there were special ethics for certain professions. In his opinion there was only one ethic, namely that of
educated people. This one ethic had to be inculcated by the family,
through appropriate upbringing of the young. It could not be taught
at university.[76]

Ziemssen responded to von Jaksch's criticism by publishing in
the following year, 1899, a booklet on medical ethics as a teaching
subject. It mainly consisted of a translation and adaptation of Jukes

de Styrap's *Code of Medical Ethics* of 1878, the unofficial code of conduct of the British Medical Association.[77] Ziemssen maintained that medical ethics was a specific subtype of applied ethics, based on Kant's categorical imperative, Schopenhauer's voice of feeling and Herbart's practical judgement.[78] However, his (respectively Styrap's) code merely elaborated upon the conventional doctors' duties to patients, to colleagues and to the state. What was at stake here was not just proof of the existence of a distinctly medical ethics, but the assumed practical usefulness of teaching ethics in view of overcrowding of the profession and an alleged loss of public reputation. Ziemssen complained that the public and the authorities did not hold doctors in sufficient esteem. The reason for this, he believed, lay in a lack of ethics of the medical profession.[79] What Ziemssen primarily meant with 'lack of ethics' was lack of collegiality. Accordingly, his text put the emphasis on intra-professional relationships, for example in consultations with other medical practitioners and specialists, or in locum practice. Teaching medical ethics at university was meant to prevent overly competitive and self-interested behaviour when the students became medical practitioners. However, the proposals to teach ethics in the medical curriculum do not seem to have taken off. A motion at the *Ärztetag* of 1902 to introduce lectures on medical ethics for medical students in their final, practical year was rejected. It was agreed to merely familiarise them generally with the tasks and duties of the medical profession, especially in implementing the regulations of the relevant social legislation.[80] Apart from Pagel in Berlin, only the Greifswald clinician Erich Peiper is known to have given student lectures specifically on medical professional ethics around this time.[81]

Moreover, the whole genre of contemporary writings on medical professional ethics was criticised by the Berlin psychiatrist and active member of the local doctors' chamber, Albert Moll. In his view this literature concentrated too much on matters of professional etiquette. Most authors, he maintained, had hitherto neglected or ignored the most important ethical issues, including refusal or breaking-off of treatment, euthanasia, the right to deceive, advising extramarital sexual intercourse, the right to abortion and the performance of cosmetic operations.[82] Moll approached the topic in a different manner, focusing instead on the ethical conflicts for doctors in the various areas of their practice. The result was his comprehensive, 650-page book *Ärztliche Ethik: Die Pflichten des Arztes in allen Beziehungen seiner Thätigkeit* (Medical Ethics: The Doctor's Duties in All Aspects of His Work), published in 1902 with Ferdinand Enke Verlag in Stuttgart. In many respects Moll's work on medical ethics reached a deeper level of discussion than those of his predecessors, and it has also been repeatedly considered in the more recent historiography of medicine and medical ethics. While Julius Henri Schulz (1986) has suggested that Moll's detailed book was influenced by the philosophical positivism of the nineteenth century and by principles of contract law, Antonia K. Eben (1998) has characterised Moll's ethical approach as inductive and casuistic, and in line with a liberal, Jewish-Christian morality.[83] Moll had in fact converted from the Jewish faith to Protestantism in 1896. Eben's and Schultz's assessments were preceded by Susanne Hahn's view that Moll's approach was that of 'situation ethics' and represented a subjective and idealist response to an increasingly science-dominated, materialistic medicine.[84] While it has remained a moot point as to how influential Moll's book actually was during its own time[85] – in contrast to other works of the genre it did not see a second edition – there can be little doubt about its conceptual significance. It will therefore be discussed in some detail in the following section, regarding its conceptual underpinnings as well as some key issues of medical ethics as identified by Moll himself.

4.5 A new point of departure?
Albert Moll's *Ärztliche Ethik* (1902)

When Moll published his *Ärztliche Ethik*, he was just about 40 years old, but already had 15 years experience as a neurologist, psychiatrist and psychotherapist in private practice in Berlin.[86] Influenced by Hippolyte Bernheim's and Ambroise-Auguste Liébeault's use of hypnosis as a form of suggestion therapy, Moll had adopted this technique for his practice, and in 1889 he had brought out his first book on this subject.[87] From the early 1890s, Moll had also published on sexology, a field in which he became an internationally recognised expert.[88] Besides his practice and scientific work, he was active in the professional politics of the medical chamber for Brandenburg–Berlin, especially in the negotiations with the sickness insurance organisations. Moll was thus both an experienced and well-placed practitioner, who could claim to make a valid contribution to the field of medical ethics. The immediate cause for his writing in this field was the contemporary debates over the use of hospital patients for clinical experimentation, including the scandal over the syphilis experiments of Albert Neisser (see Chapter 3).[89]

As Moll acknowledged in the preface to his *Ärztliche Ethik*, he had received much advice from the philosopher Max Dessoir, with whom he also collaborated on hypnosis and parapsychology within the Berlin 'Society for Experimental Psychology', and from the Berlin physician and former Breslau professor of medicine, Ottomar Rosenbach (1851–1907), who had been one of his university teachers.[90] The latter published, in 1903, a critique of the bacteriology of Robert Koch and others, arguing that they neglected individual constitutional and environmental factors in the study of infectious diseases.[91] Moll had also received support from the Berlin philosopher Georg Simmel, who was then working on sociological questions, leading in 1908 to his famous study on the 'forms of socialisation'.[92] Their intellectual input was particularly reflected in Moll's discussion of the question whether the theories of moral philosophy could provide a basis for a doctor's ethics.

According to Moll, the then current systems of moral philosophy were unsuitable as a foundation of medical ethics, for two reasons: they could be used to argue that the doctor's role as a healer was superfluous; and one and the same moral theory could be used to demonstrate that one and the same action was ethical and unethical. In particular, he took issue with the system of 'universal evolutionism' or evolutionary ethics, which regarded the progress of human society as a whole as the decisive criterion for an ethical action. Following this theory, medicine would have to refrain from treating people who have a hereditary disability, because therapy might help them to pass on their condition to the next generation. Such a eugenic and Social Darwinist perspective would thus negate the doctor's role as a healer. This, however, was completely unacceptable to Moll, because he saw the duty of care for the individual patient as the very basis of medical ethics. However, as Moll suggested with reference to Rosenbach, one could argue from the evolutionary standpoint that a disability might well play an important role for the future mental and physical development of mankind, and medical help for the disabled was thus all the more mandatory.[93]

Similarly, Moll criticised the medical implications of the philosophical utilitarianism in the tradition of Jeremy Bentham. Taking the greatest benefit of the greatest number of people as the ethical criterion, experiments on dying patients, for example bacteriological trials, might be justified in view of the potential usefulness of the results to many other patients – again something that Moll rejected from the point of view of his medical professional ethics, which focused on the health of the individual. But utilitarianism could also be employed to argue that such experiments were forbidden, because they undermined the public's trust in doctors and hospitals, leading thus to harm to a large number of people.[94]

Also moral theology, insofar as it viewed illness as punishment for sin, was unsuitable as a basis for medical ethics, according to Moll, because it could be used to argue that providing treatment might be an act against divine will.[95] Instead of moral theories or doctrines, Moll considered the 'moral feelings' of the population, an 'average morality' (*Durchschnittsmoral*), which would also be applicable to ethical issues in medicine. In particular, he proposed that it was most adequate to understand the doctor–patient relationship as a voluntary contract between two parties, with duties and rights on both sides. The contract could well be made tacitly, simply by the patient (or their legal representative) turning to the doctor for help or advice, and the doctor showing a willingness to provide it. One important implication of this concept was that the doctor was not obliged to treat any patient who requested it, that is the doctor's 'freedom to cure' (*Kurierfreiheit*) as guaranteed by the Trade Ordinance was ethically vindicated by this approach. Only emergencies formed exceptions, in which the doctor had a legal duty of assistance like any other citizen.[96]

The concept of an individual contract between doctor and patient was subsequently applied by Moll to all kinds of situations in medical practice. In this way he developed his ethical position in multiple scenarios. His book covered in detail the doctor–patient relationship (or as Moll put it, the relationship between 'doctor and client'), the various forms of medical practice (as house doctor, specialist, panel doctor, hospital doctor, etc.), ethically problematic treatments, economic aspects, professional and private conduct, the role of hygiene, work as a medical expert, medical science and research, and medical education. In the following I will outline Moll's views on the key issues of truth-telling, euthanasia and perforation of the foetus and abortion, in order to illustrate his approach. His critical position in the question of patient consent (to treatment and experimentation) has already been discussed in Chapter 3.

Moll was aware that philosophers such as Kant and Fichte had advocated truth-telling under any circumstances, but he regarded such advice – in line with his general criticism of moral theories – as unsuitable guidance for medical practice. The doctor had to distinguish between situations in which patients asked for his expert opinion on their condition and the likely success of a particular treatment, and situations in which patients entrusted themselves to the doctor's care without a specific mandate. In the first scenario, the contract relationship obliged the doctor to give the patient a truthful expert opinion. An exception could be made if the prognosis was so bad that suicide of the patient was a concern (cf. the similar view of Marx mentioned above). In the second scenario, the tacit contract demanded that the doctor acted first in the interest of the patient's health. Here, deception (e.g., by endorsing a hysterical patient's hope for a cure of paralysis through magneto-therapy) could be permissible. The probability of a treatment being successful through its suggestive effect, and the harmlessness of the method used, could justify deceiving the patient. However, deception was also justifiable in dangerous diseases, according to Moll, if the truth was withheld from the patient in her own interest. It might then even be allowed to mislead the relatives if they were likely to reveal the seriousness of the condition through their words or behaviour.[97]

In detail, Moll discussed the question of truth-telling in the case of incurable patients whose death was imminent. Based on his contract-theory of the doctor–patient relationship, he distinguished, in this context, different roles of the doctor: whether he had only been approached for his expert opinion on the prognosis; whether he was simultaneously providing the treatment; or whether he might, as the house doctor, have obligations to other family members as well as to the patient. Furthermore, he identified two reasons that might justify tactful disclosure of the dire prognosis to the patient, even if this disclosure carried a risk of leading to further deterioration of their condition: the patient's need to sort out their personal affairs, such as the drawing up of a last will; a religious (Catholic) patient's wish to be given the last rites.

If the doctor had only been asked for his opinion as a medical expert, disclosure in these two cases was relatively unproblematic. If, however, he also had to consider the effect of disclosure on the patient's condition (as the doctor in charge of the treatment) and/or on the close family (as the house doctor), he had to carefully weigh the benefits against the risk of harm. As a rule, Moll advised to involve a third person to communicate the bad prognosis, because this information would less affect the patient's psychological condition if it came from a lay person than from the doctor who was expected to bring help. Generally, Moll shared the concern of earlier writers on the subject, including Hufeland and Marx,[98] that communication of a hopeless prognosis might harm the patient (and might even trigger suicide) and should therefore be avoided under the therapeutic imperative. While his application of the contract-theory had led him to a more differentiated view of the problem, he thus did not give up the traditional medical position in this matter.[99]

In connection with his discussion of ethical behaviour in dealing with terminally ill patients, Moll also developed his stance on the issue of euthanasia. In late nineteenth-century Germany the discourse on euthanasia had begun to change, as notions of human life 'unworthy of life' started to be discussed from the perspectives of Social Darwinism, racial hygiene and eugenics. Social Darwinist authors such as the Jena zoologist Ernst Haeckel (1834–1919) and Alexander Tille (1866–1912), lecturer in German language and literature in Glasgow, had invoked the examples of infanticide in weak and malformed newborns in ancient Sparta and among contemporary North American Indian tribes and sympathised with the idea of artificial selection in man. From the standpoint of racial hygiene and eugenics, the Berlin physician Alfred Ploetz (1860–1940) had advocated in 1895 a future society that would practise euthanasia (with morphine) of weak and disabled newborns, and that would refrain from caring for the sick, the blind and the deaf-mute in order not to counteract natural selection. Six years earlier, Friedrich Nietzsche had denounced the infirm in his *Götzendämmerung* as parasites on society, who – from a certain stage onwards – should be met with social contempt. Doctors, he thought, should act as 'mediators of this contempt'.[100]

In this intellectual climate appeared, in 1895, the notorious booklet *Das Recht auf den Tod* (The Right to Death) by the Göttingen student of philosophy, mathematics and physics, Adolf Jost.[101] Although section 216 of the Reich Penal Code of 1871 determined that killing on demand had to be punished with imprisonment of not less than three years, he argued that there were cases of incurable physical or mental illness in which death was desirable both from the point of view of the patient and from the perspective of society. Taking a utilitarian approach, Jost claimed that the individual's suffering and the harm caused to society by their illness could result in a 'negative value of the human life'. On this basis he advocated legalisation of voluntary active euthanasia for such cases to be carried out by physicians. As a second step of such a 'reform' he also envisaged euthanasia without consent of incurably-ill mental patients.[102]

Moll's rejection of such Social Darwinist and eugenic proposals had already been expressed in his treatment of the general relationship between medical ethics and moral theories. In his discussion of euthanasia he made it clear that any measures which deliberately shortened the patient's life were inadmissible both from the point of view of criminal law and of morality. This was also true, according to Moll, if the patient requested them. His argument against active euthanasia was that of the slippery slope. If one admitted such a right to kill, it might be applied to shorten a patient's life for months or years, if an unproductive life full of suffering was expected. There would be no stopping. For Moll, the doctor's highest good had to be life, and death the worst evil, so his task was to prolong life, not to shorten it. Although Moll rarely referred to Hippocratic ethics, he did mention here the Hippocratic Oath's prohibition of giving a deadly poison on demand.[103] With his clear stance against active euthanasia Moll was in line with the prevailing opinion of the medical profession at this time, which had not yet been substantially affected by the Social Darwinist proposals mentioned above.[104]

Moll advocated the traditional form of euthanasia (in the sense of Marx and others) as palliative care. Pain relief was its crucial task, and narcotics should be used in sufficiently high doses, even if they induced unconsciousness. Moll was aware of the problem that the administration of strong painkillers such as morphine might shorten the patient's remaining life span. Although he was clear that deliberately life-shortening measures were forbidden, he appeared to accept this implication as part of the common practice in medicine to use treatments with a certain risk. Such risky treatments were justifiable in his view with the patient's consent. Moll did not, however, explicitly discuss the problem of indirect or double-effect euthanasia. 'Heroic' interventions or strong stimulants to prolong the patient's life for a very short time should be omitted, even if the relatives asked for them to be applied, as such measures only meant further suffering for the patient. Moll pointed out that moral practice viewed omissions less critically than acts, and he also upheld the view that the doctor's contract relationship was with the dying patient, not with the relatives. Likewise from the contract-perspective, Moll regarded experimentation on a dying person as deeply unethical and condemned it as a shameful act of brutality.[105]

His contract-perspective also helped Moll in structuring his discussion on the issue of the perforation of the foetus, that is of a craniotomy of the foetus (leading to its death) in order to save the life of the mother if a natural birth was impossible (e.g., because of too narrow a pelvis). The alternative of a Caesarean section still carried, around 1900, a very high risk of mortality. Decisive for Moll in this situation was the will of the mother. If she requested the perforation, the doctor should carry it out. The same route should be taken if the woman left the decision to the doctor or if her condition made her unable to express her will. In Moll's view the life of the foetus could not be weighed against the mother's life, because the foetus had not the status of an independent human being. If, however, the woman demanded the Caesarean section, perhaps out of a sense of self-sacrifice and being fully aware of the risks, the doctor was justified in carrying out the dangerous operation. If the mother was dying and unable to express her will due to unconsciousness, a Caesarean section to deliver a possibly living child could be performed, as her consent could be assumed.[106]

Similarly, Moll showed understanding for women who decided on an abortion in earlier stages of pregnancy, because they already had too many children to care for or if the pregnancy had resulted from rape. Penal law was in his view out of tune with public sentiment, which did not always regard abortion as something unethical, especially not if it was performed relatively shortly after conception. The health of the mother was in his opinion an even stronger moral justification for an abortion. However – and this again showed Moll's respect for the patient – if the woman rejected abortion or perforation for religious reasons regardless of the danger to her health, the doctor might try to persuade her to change her mind, but had ultimately to accept her decision.[107]

As the three examples of truth-telling, euthanasia and abortion have illustrated, Moll systematically explored ethical conflicts of medical practice under the premise of the doctor–patient relation as a contract relationship. His approach generally strengthened the self-determination of the patient. The habitus of his ideal doctor was not that of a paternalist, but that of a professional who is fully committed to the interests of his client through a (tacit) contract. However, Moll's conclusions from this approach did not necessarily put him in conflict with current medical opinion on ethical questions. His practical position on the above-mentioned issues was shared by other doctors of his time, as we have seen from the deontological writings above, although they tended to be very paternalistic. How then, did the medical profession respond to Moll's *Ärztliche Ethik*?

A friendly review in the *Münchener Medizinische Wochenschrift* by the Hamburg gynaecologist Karl Jaffé (1854–1917) asserted that Moll's book was 'a true reflection of the current views on medical ethics' and praised the author's 'impartial and objective point of view'. Jaffé thought that the text was useful both for medical students who were about to enter the profession and for experienced doctors.[108] Rather dismissive, in contrast, was the Berlin health officer, Leopold Henius. In a review for the *Deutsche Medizinische Wochenschrift* he questioned Moll's right to write a book about medical ethics in the first place, because he had recently taken a position in the issue of doctors' unrestricted admission to panel practice and patients' free access to panel doctors (*freie Arztwahl*) that differed from that of the majority in the profession.[109] More generally, Henius critically asked what the point of such a detailed volume on medical ethics was, if current opinions on proper and honourable conduct were not shared by everyone and were likely to change substantially over the next decade. Taking a similar position as von Jaksch, he claimed that those who owned a natural tactfulness did not need such a book, and those who lacked a sense of propriety and duty would hardly be improved by it. In any case, it was better to seek the personal advice of an experienced colleague in a situation of doubt than to plough through a thick handbook. Henius did acknowledge, however, the author's industriousness in covering practically all aspects of medical ethics, and he appreciated, like Jaffé, Moll's 'calm, impartial and noble tone'. Yet, Henius claimed that he had found little that was new to him in Moll's book, that it was too extensive, and that the argumentation on some issues, for example on the perforation of the foetus, was 'too sophisticated to be convincing'.[110]

Overall, Moll's *Ärztliche Ethik* seems to have met a cool reception in the medical profession. As Moll himself indicated in his memoirs, published in 1936 under National Socialist censorship, his book on medical ethics was more readily appreciated by lawyers.[111] In fact, it was regularly quoted in the legal literature on aspects of medical practice (e.g., on confidentiality and consent – see Chapters 2 and 3), which otherwise rarely cited medical authors. A very positive review in the legal journal *Der Gerichtssaal*, by its editor Melchior Stenglein, praised the similarity of Moll's position with current legal opinion on the issues of euthanasia and consent to surgery. Stenglein appreciated Moll's work as a rich source for moral decision-making in medicine and expressed trust in its author's judgement.[112] A major obstacle to the book's success as a work for medical practitioners was probably its excessive length.[113] As mentioned above, in contrast to shorter texts on medical deontology it did not go through a second edition. Moll was by far more successful as an author on hypnosis, sexology and parapsychology.[114] Despite its conceptual merits, his book on medical ethics did not become a new point of departure. Due to its unprecedented level of detail and comprehensiveness it was rather a culmination of deontological writing. How did the genre develop after Moll's contribution?

4.6 Ethics, medical law and literature on the doctor's social identity

In his recent thesis on German (and English) medical ethics literature from Moll's book in 1902 to the start of the National Socialist dictatorship in 1933, Georg Schomerus has argued that the increasingly tighter organisation of doctors vis-à-vis the sickness insurance organisations made much of the traditional writing on medical professional duties superfluous. In particular, with the foundation of the union-like *Hartmann-Bund* in 1900, which vigorously pursued German doctors' economic interests, the advice literature for incoming practitioners became less relevant as a means to mitigate intra-professional competition.[115] Another factor, which should be considered in conjunction with Schomerus' hypothesis, was that with the introduction of the medical courts of honour in Prussia and other German states the issues of professional conduct became more and more institutionalised. The published decisions of the courts of honour provided binding guidance. A further effect of the court of honour system was that formal questions of collegiality and honourable conduct continued to be prominent. Moll's plea to focus rather on the moral conflicts in the treatment of patients was not heard, or if it was, it was pointed out that professional etiquette was nevertheless important, especially for young doctors who had just entered medical practice.[116]

The changing social context affected the style and format of medical–deontological writing. While texts in the tradition of personal advice literature stayed popular to some extent,[117] works written from the perspective of the law gained prominence. A detailed handbook of medical law had been edited by the medical officers of health O. Rapmund and E. Dietrich in 1898/99, and a further one was brought out jointly by the Berlin physician Heinrich Joachim and the lawyer Alfred Korn in 1911.[118] A comparative study of medical law, considering German, Swiss, Austrian and French legislation, was published shortly before the First World War.[119] Whereas these books were essentially reference works for the medical as well as the legal profession, other authors continued to aim at the medical student or recently qualified doctor. In his introduction into medical practice of 1905, the Alsace physician K. Hundeshagen especially considered the health and social legislation and the economic interests of the doctor. The published lectures of the Greifswald professor of medicine, Erich Peiper, on 'medical professional issues' likewise considered in detail the relevant social insurance legislation and economic questions as well as the recent law on the combat of infectious diseases (1900) and its regulations on notification. Ethical questions, for example patient information and consent, were discussed by Hundeshagen and Peiper as well (see Chapter 3), but their overall framework was provided by the law.[120] Moreover, jurists with particular expertise in medical cases added to this literature. For example, Ludwig Ebermayer commented from 1911 regularly on medico-legal questions in the *Deutsche Medizinische Wochenschrift* and after the war published updated collections of the relevant legal decisions and his commentaries, covering also the doctor's liability in malpractice cases.[121]

However, what became more influential for the professional and public discourse on medical duties was a different type of literature, on the doctor's self-image, that emerged with the book *Der Arzt* (1906) by Bismarck's personal physician Ernst Schweninger (1850–1924).[122] The doctor's role and responsibility in society and his (conflicting) identities as 'healing artist' (*Heilkünstler*), scientist or racial hygienist became important topics. Disappointment about a lack of therapeutic consequences from the scientific approach to medicine, the harsh realities of the panel-doctor system, and the popularity of unorthodox methods of healing were factors in a development that after the First World War led to a wider debate about the 'crisis of medicine'. Before and after the war, several medical authors disseminated in this context their ideal conception of the doctor's identity, writing for the profession and the general public alike.[123]

As Urban Wiesing has pointed out, Schweninger's book of 1906 was a sweeping attack on the scientific medicine of the late nineteenth century and came to be seen by critical doctors of the 1920s, such as Georg Honigmann and Erwin Liek, as a turning point in the self-image of medicine.[124] Since 1900, Schweninger had been director of the District Hospital in Lichterfelde (near Berlin), where he promoted naturopathic treatments, and in 1901 he had been given a chair and teaching contract for general pathology, therapeutics and history of medicine at the Berlin university clinic, the Charité.[125] Although *Der Arzt* was probably written by his assistant Emil Klein (1873–1950) rather than by Schweninger himself, it represented the latter's opinions.[126] Schweninger rejected the view of medicine as an applied science as completely wrong and invoked instead the old image of the doctor as a 'healing artist'. The personality of the doctor was decisive; he had to be the 'ruler' in the relationship with the patient and had to treat intuitively, individually and holistically. The doctor's healing powers were god-given and motivated by his humanitarianism (*Humanität*). His 'true' art of healing was timeless, whereas medical science was subject to rapid change and 'killed' the doctor's humanitarianism.[127]

Schweninger thus endorsed the habitus of the paternalistic, sovereign doctor, who firmly led the patient – a marked contrast to Moll's conception of a contract between doctor and client. Whereas Moll's reflective and sober ethics held little appeal for the medical profession, Schweninger's idealised picture of the doctor's personality proved to be attractive, especially in the economically and politically difficult period of the Weimar Republic.[128]

4.7 Conclusions

As this discussion of the German medical ethics literature from the 1870s to the First World War has shown, doctors often engaged with key issues of their professional practice in a rather stereotypical manner. The image of the doctor as a paternalistic helper of the sick, as a skilful and discreet healer and as a man of honour vis-à-vis his patients as well as his colleagues, characterised the ideal habitus that was constructed again and again in deontological writings.

Some significant themes within this literature can be identified. The relevance of medical science for medical practitioners was, for example, contentious. The initial critique by K. F. H. Marx, that the natural sciences were overrated in medicine, especially in medical education, was muted by supporters of medical science in the 1880s and 1890s, such as the practitioners Schmidt, Mensinga and Pagel. However, a critical attitude towards excesses of medical experimentation, for example bacteriological trials on dying patients, was expressed by Moll at the turn of the century.

Another issue was whether actively life-shortening measures in terminally ill patients were morally permissible. Until (and including) Moll, the predominant stance of medical practitioners – at least on the level of their normative writings – was to reject active euthanasia and to advocate palliative care instead. Finally, on the issue of abortion doctors showed increasing understanding for women. Marx's suggestion in the 1870s to report a suspicious miscarriage to the authorities had become unacceptable to practitioners such as Wolff and Moll around the turn of the century. In general, the writers on medical ethics in the examined period developed their views from a position of paternalism. Moll's contract-based ethics, which allowed a good measure of self-determination of the patient, remained an exception that apparently found no followers in the medical profession of his time and rather appealed to legal authors.

Epilogue

The four main themes of this book – professional discipline, confidentiality, patient consent and the doctor's duties – have illustrated various dimensions of the relationship between medical ethics and the law in Imperial Germany. What were the main features of this relationship? Which continuities and discontinuities in doctors' professional ethics and in medical law can be observed after the German defeat in 1918, during the Weimar Republic?[1] Which changes came in this regard with the seizure of power by the National Socialists in 1933?

During the *Kaiserreich*, doctors had used courts of honour – an institution initially introduced in the military and in the legal profession – in order to cope with problems of intra-professional competition and overcrowding, and to deal with conflicts with the state-authorised health insurance system. The disciplinary powers of these courts were applied to enforce solidarity among medical practitioners in contracting with the boards of the sickness insurance funds. They were also employed to sharply demarcate licensed doctors from lay practitioners and to ensure that doctors were loyal to scientific, academic medicine. Links to the naturopathy movement were especially persecuted by the medical disciplinary tribunals. However, the rationale for imposing disciplinary punishments derived its force not only from the exigencies of professional politics. On another level, the disciplinary tribunals also embodied the cult of male honour which pervaded the medical profession as well as the legal profession and the officer corps. Honour, as Georg Simmel observed, gave cohesion to social groups, including the professions.

Notions of professional honour and reputation guided the decision-making of the disciplinary committees. By enforcing adherence to these notions, the medical courts of honour stabilised the profession and sought to increase its homogeneity. The medical man's 'sense of honour' (*Ehrgefühl*) made a reprimand or punishment by a disciplinary court a serious affair, as it questioned his membership in his social group as well as his personal integrity. The doctor's male honour also demanded respect for female sexual honour (*weibliche Geschlechtsehre*). Hence, the medical disciplinary tribunals punished transgressions vis-à-vis female patients as well as misconduct in the sexual sphere outside of medical practice.

In the Weimar Republic, the medical courts of honour continued their work without major changes. Also, the areas of professional discipline remained much the same.[2] A motion in the Prussian Parliament to abolish the courts of honour, proposed in 1921 by the left parties SPD, USPD and KPD had no success.[3] The disciplinary tribunals became a topic of debate at the forty-third *Ärztetag* held in 1924 in Bremen, but a professional code for German doctors, adopted two years later at the forty-fifth meeting in Eisenach, again moved the issue out of awareness. This code was actually based on a medical professional code for Prussia from 1911, which had been revised in the light of the decisions of the Prussian Court of Honour for Doctors.[4] By 1928, the doctors' chambers in nearly all German states had established medical disciplinary tribunals, whose sanctions included warnings, reprimands and fines.[5]

Only the National Socialist *Reichsärzteordnung* (Reich Doctors' Ordinance), issued on 13 December 1935, brought significant change by removing doctors from the regulations of the Trade Ordinance and by abolishing the *Kurierfreiheit*, that is the status of medical practice as a free trade. It also altered the organisation and powers of the medical conduct jurisdiction. Across the whole of the Reich, medical professional courts (*ärztliche Berufsgerichte*) were formed in the districts of the doctors' chambers. An *Ärztegerichtshof* was established in Munich to serve as a central appeal court. State-representation in these professional courts was strengthened. The medical courts at district level were each chaired by a state-appointed jurist, and two doctors served as assessors. The Reich Ministry of Justice appointed two jurists for the Munich appeal court (one of them to act as chairman), and the newly created *Reichsärztekammer* (Reich Medical Chamber)

delegated three doctors as members. Apart from warnings and reprimands, the new *Berufsgerichte* could issue fines of up to 10,000 Mark,
exclude doctors from work in the public health sector, and decide
whether they were worthy to practise medicine at all. As soon as disciplinary proceedings had been initiated, the accused doctor could
be temporarily banned from the profession.[6] The National Socialist
dictatorship thus brought a change for doctors from decentralised
professional control towards centralised control by the state.

The professional privilege of medical secrecy was likewise subor-
dinated to state interests. This tendency had already set in during
the *Kaiserreich*, when the imperatives of public health and venereal
disease prevention came to be seen as legitimate reasons for breach-
ing patient confidence in the interest of the community. In the
Weimar Republic, this position was confirmed with the legislation
on the notification of venereal diseases in 1927. Medical confidentiality was further restricted through the National Socialist Law for
the Prevention of Hereditarily Ill Offspring (*Gesetz zur Verhütung erbkranken Nachwuchses*) of 14 July 1933, which introduced compulsory
sterilisation of people who were perceived as suffering from a genetic
disease. For this purpose, it obliged all doctors to notify the health
authorities of any patient with inborn mental deficiency, schizophre-
nia, manic-depressive illness, hereditary epilepsy, Huntington's cho-
rea, blindness, deafness, severe malformations or severe alcoholism.
Section 7 of this notorious law explicitly annulled medical secrecy
in this context and required doctors to give evidence in the (non-
public) proceedings of the newly created 'hereditary health courts'
(*Erbgesundheitsgerichte*) which decided on the sterilisations. Cynically,
section 15 of the same law bound persons involved in the proceedings and in the surgical intervention to secrecy, threatening imprisonment for up to a year or a fine for unauthorised disclosure.[7]

Clearly, the eugenic drive of National Socialist health policy had
turned the legal duty of medical confidentiality into an empty shell.
The 'rationale' for sacrificing medical secrecy in the hereditary
health courts was that the state, instead of the patient, released the
doctor from his duty of confidentiality. The interest of the physically
or mentally 'inferior' individual had to make way for the collective
interest of the race.[8] In 1935, medical confidentiality was altogether
removed from section 300 of the Penal Code and inserted (in a different version) in the new Reich Doctors' Ordinance.

At first sight, the new regulation of medical secrecy in section 13 of the Reich Doctors' Ordinance appeared stricter than that of section 300. Punishment of unauthorised disclosure was increased from three months to a maximum of one-year imprisonment and the duty of confidentiality was extended to medical students and to the descendants or executors of a deceased doctor. Importantly, however, the new regulation stated that somebody who revealed a secret went unpunished if such disclosure served 'to fulfil a legal duty or a moral duty' or 'some other purpose that is justified according to the healthy judgement of the *Volk*' and if the 'threatened legal good' carried more weight.[9] The idea of a priority of public interests over individual interests had thus been fully implemented.

During the *Kaiserreich*, legal cases and public scandals about surgery without consent and dangerous clinical trials had made doctors increasingly aware that it was in their own interest to inform patients and to seek formal consent. But their medical paternalism was not seriously affected by such outward compliance with the law. As work on a penal law reform was resumed after the First World War, deliberations continued about sections that would exempt therapeutic interventions from regulations on physical injury and introduce punishment specifically only for medical treatment against the patient's will. Two draft sections (section 263 and section 281) on these matters made it – after complex negotiations and discussions – into the final draft code that was considered by the penal law commission of the *Reichstag* in 1929 and 1932. However, due to the usurpation of power by the National Socialists in 1933 and the failure of the penal law reform, none of them became law.

Similarly, the issue of consent to human experiments again became a subject of debate in the Weimar Republic. These discussions led to guidelines about the requirements of information and consent in non-therapeutic as well as therapeutic experimentation that were agreed by the *Reichsgesundheitsrat* (Reich Health Council) in 1930 and issued by the Reich Minister of the Interior in 1931.[10] The Law on the Combat of Venereal Diseases of 1927, while permitting compulsory treatment under certain circumstances, preserved a patient's right to refuse dangerous therapies, such as those with mercury preparations or with the drug Salvarsan (arsphenamine), which had been introduced in 1910 and carried considerable health risks.[11] The question of the extent of patient information about the risks of a planned therapeutic intervention then likewise became a topic of discussion. In two decisions, on 19 May 1931 and on 29 February 1932, the Supreme Court postulated a doctor's 'general duty of advice' (*allgemeine Beratungspflicht*) and a duty of truthfulness when informing patients, even if the prognosis was dire, as in cancer.[12]

Two further decisions of the Supreme Court, during the period of National Socialism, of 19 June 1936 and 8 March 1940, restated the position that medical interventions against the patient's will were illegal, because the individual's bodily integrity had to be respected. However, the tone of these decisions was defensive, as the court tried to evade the implications of the Reich Doctors' Ordinance of 1935, which emphasised the doctor's duty to the health of the *Volk* as well as his duty to the individual patient and could therefore be used to justify compulsory treatment in the interest of the public.[13] In the end, these Supreme Court decisions provided no protection whatsoever for those parts of the population that became targets and victims of persecution and annihilation under the National Socialist regime. Similarly, the Reich guidelines of 1931 on information and consent in human experimentation were blatantly disregarded or ignored during the Nazi period.

Albert Moll's case for self-determination of the patient, developed in response to dangerous human experiments in the late nineteenth century, had been marginalised, as the deontological literature of the 1920s became increasingly preoccupied with the doctor's self-image. For example, the Giessen medical historian and physician Georg Honigmann (1863–1930), taking his cue from Ernst Schweninger's *Der Arzt*, described in 1924 the doctor's personality as one of the three pillars of medicine (besides empirical knowledge and speculation) and linked it to an immortal Hippocratic, holistic mode of healing of the doctor as an artist.[14] The Danzig surgeon Erwin Liek (1878–1935), the most outspoken critic of medicine in the 'crisis' debate of the 1920s and 1930s, called Schweninger the doctor 'on whose shoulders we present doctors all stand, whether we admit it or not'.[15] In 1926, two years after its author's death, a second edition of Schweninger's book was printed, and in 1927 Liek edited Friedrich Scholz's 'conversations' on doctors and patients, appreciating their author's 'grim humour' as that of a 'true physician who sees through the miserable existence of his dear contemporaries' and yet is always willing to give a helping hand.[16]

The theme of the healing artist, whose personality and psycho-logical skills help him through the adversities of medical practice, was likewise popularised by the Munich gynaecologist and writer Max Nassauer (1869–1931), a regular contributor to the *Münchener Medizinische Wochenschrift* since 1896. In 1911, he had engaged with the discussion on a patient's right to death. In a novel about a doctor and his patient-friend, a jurist suffering from an incurable sarcoma who requests euthanasia from him, Nassauer described at length the inner moral conflicts of the doctor. Eventually the doctor decides to help his terminally ill friend and provides active euthanasia through morphine injections.[17] Particularly successful were Nassauer's later collections of entertaining, often ironic short stories about doctors and patients. Amusing as these short stories were, Nassauer also bit-terly complained in them about the 'dictatorship' of the sickness insurance organisations which strictly controlled expenditure, and he deplored that the 'medical art' was being reduced to a mere craft and business. Deeply resentful of the claim-mentality of insured patients, Nassauer saw in 1925 the health insurance legislation as a relic of 'overly social' times that had to be 'smashed'. Medical university teachers were meant to help the practitioners by going beyond the science, also teaching students the 'art of medicine' and emphasising the psychology of the sick.[18] The focus on the psycho-logical dimension of the doctor–patient relationship, linked with an advocacy of medical paternalism, was likewise apparent in the works of other authors of this time, such as the surgeons Carl Haeberlin (1878–1954) in Bad Nauheim and Albert Krecke (1863–1932) in Munich. Both emphasised the therapeutic influence of an optimistic attitude and reassuring manner of the doctor.[19] Again, the habitus of the doctor as a ruler was invoked. As Nassauer put it 'The doctor must rule over the sick from an unapproachable distance. A secret power must inhabit him, if he is meant to heal.'[20]

The various strands of thought on the doctor's dominating per-sonality and the nature of 'true medicine' as an art culminated in Erwin Liek's polemical and highly successful book *Der Arzt und seine Sendung. Gedanken eines Ketzers* (The Doctor and His Mission. Thoughts of a Heretic). First published in 1926, it went through its tenth edition in 1936. Like Moll's medical ethics, Liek's writings have repeatedly attracted attention in historical scholarship, but for entirely different reasons. Liek's views on racial hygiene led him to advocate sterilisation on eugenic grounds and to sympathise with the suggestion of euthanasia of the incurably ill. They have thus placed him among the intellectual forerunners of Nazi medicine.[21] The influential booklet on *Die Freigabe der Vernichtung lebensunwerten Lebens* (The Release of the Destruction of Life Unworthy of Living) by the Leipzig jurist Karl Binding (1841–1920) and the Freiburg psychiatrist Alfred Hoche (1865–1943) had been published in 1920. It argued for the permissibility of active euthanasia in the terminally ill (upon their request as well as in unconscious moribund patients) and in the incurably feebleminded.[22] More generally, as Heinz-Peter Schmiedebach has recently shown, the hardships and pressures on medicine caused by the First World War had led to a shift from an ethics focused on the individual patient towards a 'rationalising' attitude and collective ethics that put the interests of the community or population first.[23]

Liek, in 1926, recognised the increasing importance of biological and racial hygienist questions for the medical practitioner, but still felt that the medical consideration of the individual patient ought to have priority over these issues. Linked to a case history of an incurably ill, suicidal 87-year-old patient who had requested active euthanasia, Liek stated that his conscience did not permit him 'to destroy a life unworthy of living'. However, he acknowledged that each doctor had to follow his own conscience in such cases.[24] Moreover, in the context of his discussion of social security he branded the institutional care for feebleminded and epileptic patients as a form of 'negative selection' that promoted the 'weak and incapable' over those with 'good' hereditary dispositions, and he questioned whether it was desirable to bring up 'any genetically poorly endowed newborn'.[25]

These issues, however, were not yet very prominent in Liek's general critique of medicine.[26] His key arguments were that medicine had become overly science-oriented, that many doctors had become technology-obsessed 'medics' rather than true healers of the sick, and that the sickness and social insurance system had created an unhealthy claim-mentality in patients as well as a superficial mass-practice of profit-seeking panel doctors.[27] Against this 'mechanist-materialistic' attitude in modern medicine Liek positioned the traditional ideal of the doctor as a healing artist. His ideal doctor's habitus was that of the paternalist with a forceful and suggestive personality to whom the patient had to subject himself. The doctor always had to be the dominant, 'giving' party. Schweninger's conception of the doctor–patient relationship was quoted as exemplary in this regard. Patients, Liek claimed, sought in their doctor not only the medical expert but also something irrational – 'personality, humanitarianism, goodness, magic'. Hippocrates was praised as the archetype of a holistically thinking true doctor.[28] Health education of the public, such as in the Reich Health Week (*Reichsgesundheitswoche*) of 1926, would merely create hypochondriacs.[29] If doctors again acted as 'priests' of the healing art rather than as medics, the issue of competition from quacks would be solved.[30]

In Liek's critique of medicine, thoughts of Schweninger, Honigmann and Haeberlin (all of whom were cited in his bibliography) were blended and elaborated. In contrast, Moll's contract-based medical ethics was no longer relevant and not cited by Liek. As a doctor of Jewish descent, Moll eventually lost his practising licence in 1938, although he had joined the nationalist *Deutsche Vaterlandspartei* in 1917. The harder the economic circumstances for medical practitioners became, it seems, the more authoritarian became their ideas of the doctor–patient relationship. The frustrations of doctors through a powerful health insurance system found relief in exaggerated ideas of the doctor as sovereign, ruler, giver and psychologically superior artist in the relationship with their patients. In this sense Liek's work contributed to the spread of an authoritarian conception of the doctor–patient relationship.[31] The dangers of such self-elevation were soon to become all too apparent in the abuses of medicine during the Nazi period – from involuntary euthanasia of the disabled and mentally ill to terminal human experimentation.[32]

In the context of the debate of a German special path (*Sonderweg*), historian Michael H. Kater has argued that the professionalisation and socialisation of medical students and doctors in the *Kaiserreich* and the Weimar Republic led them to adopt authoritarian, militaristic, anti-Semitic and anti-feminist attitudes, which predisposed them to their far-reaching affiliation with Nazism.[33] As my study has shown, the medical ethics of Imperial Germany were indeed not founded on deeply held moral convictions, but had been built around various responses of a male professional honour to challenges posed by the market for medical services and by the law. Compatibility with the doctor's honour and professional reputation rather than patients' interests or wishes determined medical decision-making.

As it became dramatically clear during the Third Reich, this reactive, paternalist and conservative type of professional ethics was too weak to prevent the atrocities of Nazi medicine.[34]

However, it would be rash to conclude that the paternalist and doctor-centred medical ethics of the *Kaiserreich*, as described in this book, can serve as an argument in favour of the German *Sonderweg* thesis. As the findings of recent studies into the history of British medical ethics suggest, there may have been more similarities than differences between the conduct of German and British doctors around 1900. Like the German medical courts of honour, the General Medical Council (GMC) was preoccupied with issues of professional competition, such as advertising, canvassing, covering unqualified assistants and unregistered practice, in addition to sexual offences, drunkenness and the performance of abortions.[35] Moreover, the British notion of the doctor as a gentleman was rather close to that of the German doctor as a man of honour (*Ehrenmann*). As the English physician Jukes de Styrap stated in his widely accepted (though unofficial) *Code of Medical Ethics* (1878, 4th edition 1895), a member of the medical profession had the duty 'to exert his abilities to promote its honour and dignity, to elevate its status, and extend its influence and usefulness'.[36] Any businesslike behaviour, such as advertising, agreements with druggists or the dispensing of secret remedies, were considered unworthy of a British gentleman doctor.[37] Furthermore, like their German counterparts, British doctors sought to defend their privilege of medical secrecy against interests of the state and the legal system;[38] and like some of their German colleagues, some British surgeons were accused of having operated without sufficient information and consent of their patients.[39] Medical ethics in Imperial Germany probably was, we may conclude from this, less unique than has been hitherto assumed. However, any more definite conclusion regarding the question of a German *Sonderweg* in doctors' ethics will have to await detailed comparative historical studies on all the relevant issues of medical ethics in late nineteenth- and early twentieth-century Germany and other European countries. Hopefully, this book may stimulate others to undertake such further research.

Notes

Introduction

1. Cf. Moll 1902, p. 1. All translations of German quotations are mine.
2. For medical ethics and the medical profession during the First World War, see Schmiedebach 2001; Wolff 1997; Eckart and Gradmann 1996.
3. Weindling 1989, p. 10.
4. Elkeles 1996a.
5. Brand 1977. On the impact of the concept of professionalisation on medical historiography see Burnham 1998.
6. Huerkamp 1985. For a concise English summary of this book see Huerkamp 1990.
7. Drees 1988; Jütte 1997a; Berg and Cocks 1997.
8. On the development of the British and American medical profession see for example Berlant 1975; Waddington 1984; Bartrip 1996; Baker, Caplan, Emanuel and Latham 1999.
9. Labisch 1997, pp. 41–42.
10. Herold-Schmidt 1997, p. 83.
11. See also Maehle 1999. For the tribunals' activities during the Weimar Republic see Rabi 2002; and on the specific problem of medical advertising see Binder 2000.
12. It complements in this respect the work of Nye 1995 on honour codes in the French medical profession and of Morrice 2002a on the professional ethics of the British Medical Association in the late nineteenth and early twentieth centuries. See also Nye 1997. On the disciplinary functions of the British General Medical Council see Smith 1994.
13. See also Sauerteig 1999; Maehle 2003a.
14. See also Prüll and Sinn 2002.
15. Cf. Maehle 1990; Wolff 1996.
16. See also Schomerus 2001.

Chapter 1

1. Stolberg 1998, pp. 71–72. On the concept of 'profession' in medical historiography see Burnham 1998.
2. Frevert 1995, pp. 28–84; McAleer 1994, pp. 86–103.
3. See Thilo 1864; Hochtritt 1969, pp. 29–31; Frevert 1995, pp. 129–130; Siegrist 1990, pp. 58–60.
4. For Prussia see also Huerkamp 1985, pp. 254–272; Maehle 1999.
5. Rapmund and Dietrich 1898/99, p. 58.
6. Huerkamp 1985, pp. 249–251.

7. For discussions of this concept and its practical implications in eighteenth-century society, see Wahrig and Sohn 2003.
8. Eulenburg 1896; Huerkamp 1985, pp. 242–243.
9. Rapmund and Dietrich 1898/99, p. 68.
10. See Weindling 1990.
11. See also Spree 1989, who has argued that German doctors' fight against quacks (*Kurpfuscher*) in the late nineteenth century served as a standardisation of the medical profession as well as the suppression of lay healers and self-medication.
12. Eulenburg 1896; Huerkamp 1985, p. 258; Herold-Schmidt 1997, p. 55. A typical example of the dissatisfaction with the Trade Ordinance was an application of two medical district societies (Ingolstadt-Pfaffenhofen and Aichach-Schrobenhausen-Friedberg) to the doctors' chamber of Upper Bavaria to petition the Bavarian government for effecting changes to the Ordinance in order to restrict its applicability to doctors and to punish non-licensed healers, medical assistant personnel and midwives who had encroached upon the territory of medical treatments. The application was motivated with the need to maintain the 'dignity of the medical profession' and to suppress 'the joint quackery of lay people and medical assistant personnel'. The chamber unanimously accepted the application. Cf. Protokoll der Sitzung der Aerztekammer von Oberbayern, München, 30 October 1878, pp. 4–6, in Staatsarchiv München (StAM) RA Nr. 57094.
13. Hochtritt 1969, pp. 30–33; McClelland 1991, pp. 87–91.
14. Graf 1880; Gabriel 1919, p. 30.
15. See Jütte 1997b.
16. Graf had been chairman of the Düsseldorf medical society (*Verein der Ärzte des Regierungsbezirks Düsseldorf*) since 1867 and founder of a regional society for the promotion of public health (*Niederrheinischer Verein für öffentliche Gesundheitspflege*) in 1869. In 1873 he was elected as chairman of the newly founded German Association of Doctors' Societies. See Wallichs 1895b; Tüllmann 1938; Herold-Schmidt 1997, p. 48.
17. Cf. Graf 1880, col. 49.
18. Ibid., cols 49–50.
19. Coblenzer ärztlicher Bezirksverein 1880, cols 129–134.
20. See Gabriel 1919, pp. 21–27, 37; Herold-Schmidt 1997, p. 53.
21. Hoffmann 1880, col. 234.
22. Ibid., cols 236–237.
23. Ibid., col. 238.
24. Runge 1881, cols 7–8.
25. Mestrum 1881, cols 170–178.
26. Ärztetag 1881, cols 193–199.
27. Ärztetag 1882, cols 209–217; Gabriel 1919, pp. 60–63.
28. Gabriel 1919, pp. 63–64.
29. Ibid., pp. 64–65. The revised Trade Ordinance of 1 July 1883 confirmed, however, the status of medical doctors as traders (*Gewerbetreibende*) and

the freedom to practise medicine for anyone, restricting only the use of the title *Arzt* and appointments to official medical positions and as vaccinators to licensed (*approbierte*) medical practitioners. See Pistor 1890, pp. 11–12.

30. Guttstadt 1890, pp. 51–54; Gabriel 1919, pp. 77–79; Huerkamp 1985, pp. 262–263.
31. Cf. Königliche Verordnung, betreffend die Einrichtung einer ärztlichen Standesvertretung. Vom 25. Mai 1887, reprinted in Guttstadt 1890, pp. 54–59.
32. Between 1876 and 1887 the number of doctors in Prussia had increased from 7,963 to 9,284. This increase of 16.6 per cent was for the first time faster than the population growth in Prussia, which was 10 per cent between 1875 and 1885; cf. Huerkamp 1985, p. 150.
33. Gabriel 1919, pp. 82–83, 94.
34. Ibid., pp. 94–95. Bismarck's letter of 3 May 1889 was printed in *Ärztliches Vereinsblatt für Deutschland* 16 (1889), col. 155.
35. In fact, such regulations for the whole Reich were only introduced under the National Socialist dictatorship through the *Ärzteordnung* of 1935, which was linked to Hitler's programme of forcing into line (*Gleichschaltung*) of institutions, to the abolition of the *Kurierfreiheit*, and the removal of the medical profession from the Trade Ordinance. See Knüpling 1965; Wolff 1997, pp. 119–123; Rüther 1997, pp. 173–175.
36. Ärztetag 1889, cols 273–291, 321–340. On the issue of the trade with secret remedies (*Geheimmittel*), see Woycke 1992; Wimmer 1992; Wimmer 1994.
37. See Ärztlicher Bezirksverein München 1875; Altmann 1900, pp. 179–183.
38. See Baker 1995a.
39. Ärztetag 1889, cols 273, 334.
40. Cf. Graf 1890, pp. 138–160. For the Kingdom of Prussia, Graf listed 184 medical societies, of which 68 had a professional code and 92 a disciplinary tribunal.
41. Cf. Gabriel 1919, pp. 98, 100.
42. Geheimes Staatsarchiv Preußischer Kulturbesitz (GStAPK) I. HA Rep. 84a, Nr. 101, Bl. 56–63; see also Joachim [1913], p. 40.
43. Gabriel 1919, pp. 101–104.
44. The text of the instruction is quoted in *Ärztliches Vereinsblatt für Deutschland* 19 (1892), col. 43; see also Gabriel 1919, pp. 104–105.
45. See Ausschuss der preussischen Ärztekammern 1892.
46. Cf. Erweiterung der Disciplinarbefugnisse der preussischen Aerztekammern, *Ärztliches Vereinsblatt für Deutschland* 19 (1892), cols 86–92; Ueber einige im Reichstag und preussischen Landtag verhandelte Fragen, ibid., cols 117–121; Aus dem stenographischen Bericht des preussischen Landtages, ibid., cols 135–143; Cnyrim 1892; Gaertner 1896, pp. 35–36.
47. Cf. Ausschuss der preussischen Ärztekammern 1892, col. 415. Bosse's response is quoted in Die Disciplinarbefugnis der preussischen Aerztekammern, *Ärztliches Vereinsblatt für Deutschland* 20 (1893), cols 205–207.

48. The advert, originally published in the *Ärztlicher Central-Anzeiger* by Dr Landmann in Boppard, is quoted in Die Krankenkassen und die Socialdemocratie, *Ärztliches Vereinsblatt für Deutschland* 21(1894), cols 571–572.

49. Cf. Aus den preussischen Aerztekammern, ibid., 22 (1895), cols 123–124.

50. See Gabriel 1919, pp. 121–122.

51. GStAPK I. HA Rep. 76 VIII B, Nr. 770, Bl. 64–65.

52. Entwurf eines Gesetzes, betreffend die ärztlichen Ehrengerichte, das Umlagerecht und die Kassen der Aerztekammern, *Ärztliches Vereinsblatt für Deutschland* 23 (1896), cols 193–194, 227–233.

53. See Gabriel 1919, pp. 124–138.

54. Cf. Entstehungsgeschichte des § 3 Abs. 3 des Gesetzentwurfs, betreffend die ärztlichen Ehrengerichte, das Umlagerecht und die Kassen der Aerztekammern, report by Konsistorial-Assessor Altmann to the Minister, dated 1 November 1899, GStAPK I. HA Rep. 76 VIII B, Nr. 771, Bl. 246–251v. See also Luther 1975, pp. 21–22.

55. National-Zeitung, 29 March 1896, in GStAPK I. HA Rep. 76 VIII B, Nr. 784.

56. Gesetz, betreffend die ärztlichen Ehrengerichte, das Umlagerecht und die Kassen der Aerztekammern. Vom 25. November 1899 (Preuß. Gesetz-Sammlung, S. 565). Also printed in *Ärztliches Vereinsblatt für Deutschland* 27 (1900), cols 3–11. Commentaries on this law are Altmann 1900 and Kade 1906.

57. See Hochtritt 1969, p. 31.

58. See GStAPK I. HA Rep. 76 VIII B, Nr. 793.

59. Altmann 1900, p. 40; Kade 1906, p. 42; Preußischer Ehrengerichtshof für Ärzte 1908–1934, vol. 1, pp. 164–165 (decision of 29 November 1901).

60. This was a substantial sum, given that in 1900 approximately 30 per cent of the doctors in Berlin–Brandenburg had an annual income of less than 3,000 Mark, and approximately 17 per cent had an income between 3,000 and 5,000 Mark p.a. Cf. Huerkamp 1985, p. 214.

61. Altmann 1900, p. 84.

62. Before his career in the Ministry (up to *Ministerialdirektor und Wirklicher Geheimer Oberregierungsrat*), Förster had been a judge. See Sauerteig 1999, p. 509; Weindling 1990, p. 207.

63. See the obituary by Wallichs 1895a and the short biography by Wallichs 1895b.

64. See the stenographic report on the first debate of the medical courts of honour bill, 4 February 1899, Abgeordnetenhaus, in GStAPK I. HA Rep. 84a, Nr. 101, Bl. 173–182v.

65. GStAPK I. HA Rep. 76 VIII B, Nr. 771, Bl. 251–251v.

66. Standesordnung und Ehrengerichtsordnung für die ärztlichen Bezirksvereine, in Sächsisches Hauptstaatsarchiv (SächsHStA) Min. des Innern, Nr. 15146; Aerztliche Ehrengerichtsordnung, in SächsHStA Min. des Innern, Nr. 15147. For an account of the development of courts of honour in Saxony see Gabriel 1919, pp. 198–252. See also Rabi 2002, pp. 44–45.

67. See the minutes of the medical chamber meetings for Upper Bavaria 1872–1892 in Staatsarchiv München (StAM) RA, Nr. 57093–57095; and for all Bavaria 1872–1909 in Bayerisches Hauptstaatsarchiv (BayHStA) MInn, Nr. 61397–61428.
68. Cf. Gabriel 1919, pp. 167–171. See also Gutachten des Kreismedizinalausschusses von Oberbayern, erstattet in der Sitzung vom 4. Oktober 1894, in BayHStA MInn, Nr. 61413.
69. Vorschläge für eine gleichmässige Geschäftsordnung der 8 Aerztekammern. (Siehe K. A. V. v. 9. VII. 1895, § 6, Abs. 2), in BayHStA MInn, Nr. 61414.
70. Vorschläge für diejenigen Bestimmungen, welche den einzelnen Bezirksvereinen zur gleichmässigen Aufnahme in ihre Statuten durch die Aerztekammer empfohlen werden sollen, in BayHStA MInn, Nr. 61414.
71. Protokoll der XXIV. ordentlichen Sitzung der Ärztekammer von Unterfranken und Aschaffenburg, Würzburg, 29. Oktober 1895, in BayHStA MInn, Nr. 61414.
72. Geschäftsordung der Aerztekammern, in BayHStA MInn, Nr. 61415.
73. See for example a complaint discussed by the medical chamber of Lower Franconia and Aschaffenburg, Protokoll der XXIV. ordentlichen Sitzung der Ärztekammer von Unterfranken und Aschaffenburg, Würzburg, 29. Oktober 1895, in BayHStA MInn, Nr. 61414.
74. Gabriel 1919, p. 172. The same problem was also evident in disputes between doctors in other parts of the Reich, for example in a case in Thuringia in 1899. If, as in this case, the doctor accused of professional misconduct was not a member of the local medical society, this society had no means to bring him in front of its court of honour; see 'Dr. Pauling und der ärztliche Verein von Pössneck etc', *Ärztliches Vereinsblatt für Deutschland* 27 (1900), cols 171–175.
75. Cf. Protokoll der Sitzung der Aerztekammer der Oberpfalz und von Regensburg, Regensburg, 29. Oktober 1896, in BayHStA MInn, Nr. 61415; Protokoll der Situng der Aerztekammer von Schwaben und Neuburg, Augsburg, 29. Oktober 1896, ibid., S. 19–21; preamble to Entwurf einer Standesordnung für die Aerzte Bayerns, in BayHStA MInn, Nr. 61416.
76. See Entwurf einer Standesordnung für die Aerzte Bayerns, and attached report by G. Merkel on the meeting of the chairmen of the permanent commissions of the Bavarian medical chambers, held in Nuremberg, 1897, in BayHStA MInn, Nr. 61416.
77. See for example Protokoll der Sitzung der Ärztekammer von Unterfranken, Nürnberg, 30. Oktober 1897, recording unanimous approval; and Protokoll der Sitzung der Ärztekammer von Schwaben und Neuburg, Augsburg, 30. Oktober 1897, recording the proposal to follow the model of the Lawyers' Ordinance; both in BayHStA MInn, Nr. 61416. On the 1898 proceedings of the Bavarian medical chambers see the summary by Brauser 1899. See also Gabriel 1919, pp. 173–176.
78. Die bayerischen Aerzte und ihre vorgeschlagene Standesordnung, *Münchener Post und Augsburger Volkszeitung* 11, Nr. 281, 12./13. Dezember 1897, in BayHStA MInn, Nr. 61416.

79. See Stenographischer Bericht über die Verhandlungen der bayerischen Kammer der Abgeordneten, Bd. XI, 391. Sitzung, 25. Februar 1898 (pp. 321–325), and 392. Sitzung, 26. Februar 1898, in BayHStA MInn, Nr. 61416.
80. Cf. ibid., p. 321.
81. Ibid., pp. 321–322.
82. Ibid., p. 325.
83. Cf. Entwurf einer Standesordnung für die Aerzte Bayerns nach den Ergebnissen der Berathung des verstärkten k. Obermedizinalausschusses vom 19. Dezember 1898, in BayHStA MJu, Nr. 15367.
84. Cf. Letter of Minister of the Interior von Feilitzsch to the Minister of Justice von Leonrod dated 11 July 1899, asking for quick treatment of the matter in time for the next session of the *Landtag*, and reply with commentaries from the Ministry of Justice, dated 28 July 1899, in BayHStA MJu, Nr. 15367.
85. Entwurf eines Gesetzes, die ärztliche Standes – und Ehrengerichtsordnung betreffend, *Beilagen zu den Verhandlungen der Kammer der Abgeordneten* 1899, Bd. I, Beilage 10, pp. 354–358, in BayHStA MJu, Nr. 15367.
86. Cf. ibid., p. 357.
87. Ibid., pp. 356–358; see also Heinze 1899. For Saxony: Standesordnung und Ehrengerichtsordnung für die ärztlichen Bezirksvereine, II. Ehrengerichtsordnung, § 3, SächsHStA Min. des Innern, Nr. 15146, Bl. 127v.
88. *Verhandlungen der Kammer der Abgeordneten*, 16. Sitzung vom 30. Oktober 1899, p. 373, in BayHStA MJu, Nr. 15367.
89. Cf. Vom bayerischen Landtag. (Aerztliche Standes – und Ehrengerichtsordnung.), *Ärztliches Vereinsblatt für Deutschland* 27 (1900), col. 11. See also Gabriel 1919, p. 182.
90. See the obituary by Dr Wille (Bezirksarzt in Markt Oberdorf, Allgäu, und Vorsitzender des ärztlichen Bezirksvereins Allgäu), In Memoriam, off-print from *Bayerisches Aerztliches Correspondenz-Blatt*, 1900, in BayHStA MInn, Nr. 60399.
91. Cf. Die zukünftige ärztliche Standes- und Ehrengerichtsordnung in Bayern, *Ärztliches Vereinsblatt für Deutschland* 28 (1901), cols 516–518; and Gabriel 1919, pp. 184–188.
92. Cf. Protokoll der Sitzung der Pfälzischen Aerztekammer zu Speyer, 28. Oktober 1901, p. 4, Protokoll der Sitzung der Aerztekammer von Oberbayern, München, 28. Oktober 1901, p. 3, Protokoll der Sitzung der Aerztekammer für Mittelfranken, Ansbach, 28. Oktober 1901, pp. 66–67, all in BayHStA MInn, Nr. 61420; Protokoll der Sitzung der Aertzekammer von Oberbayern, München, 27. Oktober 1902, pp. 4–6, in BayHStA MInn, Nr. 61421.
93. Gabriel 1919, p. 188.
94. Cf. Entwurf der bayerischen ärztlichen Standesordnung, *Ärztliches Vereinsblatt für Deutschland* 31 (1904), cols 329–330.
95. See Mayer 1908; also reprinted within the article: Zur Reform der ärztlichen Ehrengerichte in Bayern, *Ärztliches Vereinsblatt für Deutschland* 35 (1908), cols 708–710.

96. Cf. Tagesgeschichtliche Notizen, *Münchener Medizinische Wochenschrift* 54 (1907), pp. 758–759, 2462, 2509. See also letter of Ministry of the Interior to Ministry of Justice, dated 16 January 1912, in BayHStA MJu 15393; and Doerfler 1908.

97. Mayer 1908, p. 1789.

98. Doerfler 1908.

99. Mayer 1908, p. 1789. See also Gabriel 1919, p. 190.

100. See the minutes of the meetings of the 8 Bavarian medical chambers in the years 1908 and 1909 in BayHStA MInn, Nr. 61427 and 61248.

101. Cf. Gabriel 1919, pp. 192–194. The reference to membership in an 'economic organisation' very probably referred to the Association for the Doctors of Germany for the Protection of their Economic Interests, the so-called *Hartmann-Bund* (founded in 1900 by the Leipzig physician Hermann Hartmann), which represented doctors in the disputes with the sickness insurance boards and organised 'doctors' strikes'. By 1911, 76.6 per cent of the civil doctors in the German Reich were members of this association. Cf. Herold-Schmidt 1997, pp. 50–51, 90–94.

102. Gabriel 1919, p. 197.

103. Cf. Entwurf eines Gesetzes über die Standesvertretung, die Standespflichten und das ehrengerichtliche Verfahren der Ärzte, in BayHStA MJu, Nr. 15367.

104. See BayHStA MJu, Nr. 15393.

105. Finkenrath 1928, pp. 51–52; Herold-Schmidt 1997, p. 53.

106. Cf. Doerfler 1908, p. 1997; similarly the anonymous commentary in *Ärztliches Vereinsblatt für Deutschland* 28 (1901), cols 516–518, and the 'Tagesgeschichtliche Notizen', *Münchener Medizinische Wochenschrift* 1908, no. 34, p. 1814.

107. See Preußischer Ehrengerichtshof 1908–34.

108. GStAPK I. HA Rep. 76 VIII B, Nr. 782–783. For the activities of the Prussian and Saxon medical courts of honour during the Weimar Republic, see Rabi 2002.

109. See Gesetz über die Gewährung von Straffreiheit und Strafmilderung bei ehrengerichtlichen Strafen und ehrengerichtlichen Verfahren gegen Ärzte vom 14. Juli 1919 (Gesetz-Sammlung, S. 117), and the relevant instruction, Ausführungsanweisung, in *Ministerial-Blatt für Medizinalangelegenheiten* 19 (1919), pp. 187, 191–192. Some amnesty cases were already dealt with during the war, see GStAPK I. HA 76 VIII B, Nr. 594 (January 1915 to March 1916).

110. See Medizinal-Abteilung des Ministeriums 1905–15.

111. Cf. Preußischer Ehrengerichtshof 1901 and 1903a. See also Preußischer Ehrengerichtshof 1908–34, vol. 1, pp. 17, 170–171. For a full discussion of medical advertising see Binder 2000.

112. Cf. Preußischer Ehrengerichtshof 1903b. See also Preußischer Ehrengerichtshof 1908–34, vol. 1, pp. 81–83.

113. Ibid., vol. 1., pp. 137–145. In the Kingdom of Saxony the medical courts of honour followed the same rigorous line regarding associations

between licensed doctors and lay naturopaths, see Heinze 1901 and [Anon.] 1903a.

114. For histories of the German health insurance schemes, their relationship with the medical profession and social impact see for example Frevert 1984; Huerkamp 1985; Göckenjan 1985. A useful overview in English has been given by Labisch 1997.

115. Cf. also the compilation of reasons for disciplinary convictions by the medical court of honour for Brandenburg–Berlin in Joachim [1913], pp. 49–52. See also Winkelmann 1964.

116. See for example Preußischer Ehrengerichtshof 1908–34, vol. 3, pp. 193–206. The difficult relationship between the medical profession and the health insurance boards was also a frequent topic of the annual meetings (*Ärztetage*) of the German Association of Doctors' Societies; see the compilation for the years 1890 to 1912 by Heinze 1918.

117. See Preußischer Ehrengerichtshof 1908–34, vol. 1, pp. 106–111, 147–150.

118. See also Kade 1906, pp. 55–84 and Kaestner 1911.

119. For a full discussion of issues of medical confidentiality see Chapter 2.

120. A similar spectrum of disciplinary offences can be identified from surviving records (1911–1921) of the Saxon appeal court for medical disciplinary proceedings, the *Ärztlicher Ehrengerichtshof* in Dresden. See SächsHStA Min. des Innern, Nr. 15199. The similarity between the Prussian and Saxon key issues in disciplinary cases, with professional problems dominating over problems of the doctor–patient relationship, has also been identified for the period of the Weimar Republic (1918–1933) by Rabi 2002, pp. 54–57. On the legal history of medical malpractice, see Riegger 2007.

121. Cf. Luther 1975, p. 24.

122. Cf. Joachim 1902 and 1903; Preußischer Ehrengerichtshof 1908–34, vol. 1, pp. 112–113.

123. Cf. Preuß.Ehrengerichtshof 1908–34, vol. 1, pp. 165–169; Preußischer Ehrengerichtshof 1907 and 1910.

124. Preußischer Ehrengerichtshof 1908–34, vol. 1, pp. 141–145, 166, 168. See also Regin 1995, pp. 359–363, whose book provides a detailed discussion of the difficult relationship between naturopathy and the medical profession in Imperial Germany. For the wider context of protest movements against orthodox medicine in that period, including also anti-vaccination, anti-vivisection and anti-psychiatry movements, see Dinges 1996.

125. Kaestner 1911, p. 353.

126. Cf. GStAPK I. HA Rep. 76 VIII B, Nr. 591, Bl. 84–86.

127. Cf. GStAPK I. HA Rep. 76 VIII B, Nr. 592, Bl. 149–155. See also Preußischer Ehrengerichtshof 1908–34, vol. 2, pp. 10–11.

128. Cf. Urteile vom 13. Oktober, *Ärztliches Vereinsblatt für Deutschland* 31 (1904), col. 92. See also Preußischer Ehrengerichtshof 1908–34, vol. 1, pp. 102–103.

129. Cf. Mulert 1904.

130. Köppen 1901. On the contemporary discussion about medical ethics as a teaching subject see Chapter 4.

131. For a discussion of Simmel's conception of honour see Vogt 1997, pp. 153–186. For introduction into Simmel's sociological and philosophical work in general, see Featherstone 1991 and Frisby 2002.
132. Cf. Simmel 1958, p. 403.
133. Ibid., pp. 326, 404–405.
134. Binding 1890, pp. 27–28.
135. Simmel 1958, p. 326.
136. Cf. the interpretation and discussion of Bourdieu's theory of capitals and of honour as symbolic capital in Vogt 1997, pp. 104–152; see also Robbins 2000, pp. 32–37; Martin and Szelény 2000. The central original text by Bourdieu on his concept of honour as symbolic capital is Bourdieu 1972.
137. See also Rabi 2002, p. 220.
138. Binding 1890, p. 25.
139. Cf. ibid., p. 39.
140. See Schnitzler 2002. The story was first published on 25 December 1900 in the Christmas issue of the *Neue Freie Presse*.
141. On the ideal doctor's habitus as transported by the contemporary literature on medical ethics, see Chapter 4.
142. Cf. Kammer der Abgeordneten, 392. Sitzung vom 26. Februar 1898, in BayHStA MInn, Nr. 61416.
143. Gaertner 1896, p. 20.

Chapter 2

1. 'What I may see or hear in the course of the treatment or even outside of the treatment in regard to the life of men, which on no account one must spread abroad, I will keep to myself holding such things shameful to be spoken about.' From the Hippocratic *Oath*, translation by Ludwig Edelstein, in Edelstein 1967, p. 6.
2. Placzek 1909, pp. 2–3.
3. Ibid., pp. 49–52. For discussions on the development and implications of the French law on medical secrecy, see Villey 1986 and Tod 1979.
4. Placzek 1909, pp. 4–7; Schumacher 1931, pp. 8–10.
5. *Stenographische Berichte über die Verhandlungen des Reichstages des Norddeutschen Bundes. I. Legislatur-Periode. – Session 1870*, vol. 2 (1870), pp. 732–733 (37. Sitzung, 7. April 1870).
6. § 300 *RStGB*: 'Rechtsanwälte, Advokaten, Notare, Verteidiger in Strafsachen, Ärzte, Wundärzte, Hebammen, Apotheker, sowie die Gehülfen dieser Personen werden, wenn sie unbefugt Privatgeheimnisse offenbaren, die ihnen kraft ihres Amtes, Standes oder Gewerbes anvertraut sind, mit Geldstrafe bis zu 1500 Mark oder mit Gefängnis bis zu 3 Monaten bestraft. Die Verfolgung tritt nur auf Antrag ein.'
7. While I have been unable to find statistics on convictions of doctors for breach of section 300, fines rather than imprisonment appear to have been the common penalty. See also Maehle 2003a; Pranghofer and Maehle 2006.

8. Cf. Grassmann 1902, col. 407.
9. Schumacher 1931, p. 69.
10. Krahmer 1875.
11. Marcus 1884.
12. Auerbach 1902.
13. Liebmann 1886, pp. 7–9; Kohler 1904, p. 30; Exner 1909, pp. 43–45.
14. Reichsgericht 1889; Olshausen 1892, p. 1142; Fromme 1902, pp. 19–22; Gross 1903, pp. 246–247; Zschok 1903, p. 46.
15. Placzek 1893, p. 7; Moll 1902, p. 93.
16. Simonson 1904; Placzek 1909, p. 38; Schumacher 1931, p. 86.
17. Reichsgericht 1903; Biberfeld 1902. See also the account given by the doctor concerned: H. 1903.
18. Placzek 1893, p. 6; Exner 1909, pp. 40–41.
19. See, however, a case from 1898 in Pößneck, where legal proceedings were initiated against a doctor who refused to disclose to the judge whether he had treated on a particular day a person with a hand injury. A gas pipe had been damaged in the court on that day, leading to the danger of an explosion, and the police had questioned all doctors in town in order to obtain hints on a suspect. All the other doctors had answered the question. Cf. Placzek 1909, pp. 145–147.
20. [Anon.] 1903b; Placzek 1909, pp. 141–142.
21. *Stenographische Berichte über die Verhandlungen des Reichstags, XI. Legislaturperiode, I. Session, erster Sessionsabschnitt, 1903/1904*, vol. 2 (1904), pp. 1417–1418 (46. Sitzung, 1. März 1904).
22. Litten 1904/05. See also Krauß 1904/05a.
23. *Stenographische Berichte über die Verhandlungen des Reichstags, XI. Legislaturperiode, I. Session, erster Sessionsabschnitt, 1903/1904*, vol. 2 (1904), p. 1420. On the penal law reform see also Chapter 3.
24. For reports of these workers' uprisings see for example 'The Rioting in Berlin', *The Times*, 30 September 1910, p. 5.
25. Hellwig 1910.
26. Ebermayer 1910.
27. See for example Chotzen and Simonson 1903/04.
28. Cf. Reichsgericht 1903, p. 317 (emphasis as in the original German).
29. For example Rudeck 1902, p. 10; Moll 1902, p. 109; Placzek 1909, pp. 104–105.
30. Scholz 1914, p. 118.
31. Cf. Ottmer 1902.
32. *Es werde Licht!* (Director: R. Oswald), Berlin, Richard Oswald-Film GmbH, 1917; *Dürfen wir schweigen?* (Director: R. Oswald), Berlin, Nero-Film GmbH, 1926.
33. 'Durch die Allerhöchste Cabinets-Ordre vom 8. August 1835 genehmigtes Regulativ über die sanitäts-polizeilichen Vorschriften bei den am häufigsten vorkommenden ansteckenden Krankheiten', in Augustin 1838, pp. 29, 978–979.
34. Placzek 1909, pp. 157, 159–160; Sauerteig 1999, pp. 320–321.
35. See Evans 1987.

36. Cf. 'Entwurf eines Gesetzes, betreffend die Bekämpfung gemeingefähr-licher Krankheiten', in *Stenographische Berichte des Deutschen Reichstages, 10. Legislaturperiode, 1. Session, 1898–1900*, supplement vol. 6, p. 4189.
37. Ibid., p. 4215; Dittenberger 1906; Exner 1909, pp. 49–50; Placzek 1909, p. 159; Sauerteig 1999, pp. 326–327.
38. See *Stenographische Berichte über die Verhandlungen des Preußischen Hauses der Abgeordneten, 19. Legislaturperiode, V. Session, 1903*, vol. 2, cols 1443–1474; *20. Legislaturperiode, I. Session 1904/05*, vol. 1, cols 308–334, vol. 4, cols 5923–5945, vol. 6, cols 8530–9324, vol. 8, cols 12641–12670, 12673–12694, 12699–12728.
39. Cf. Sauerteig 1999, p. 322. Sauerteig's book provides a detailed discussion of the politics of the fight against venereal diseases in Imperial Germany and the Weimar Republic.
40. For an analysis of this society, the *Deutsche Gesellschaft zur Bekämpfung der Geschlechtskrankheiten*, see Sauerteig 1999, pp. 89–125.
41. Neisser 1905, pp. 5–26.
42. Bernstein 1905.
43. Flesch 1905.
44. The minutes of the discussion were printed in *Zeitschrift für Bekämpfung der Geschlechtskrankheiten* 4 (1905), pp. 52–72. Ludwig Quidde used the opportunity to renew his accusation against Neisser of having experimented irresponsibly with syphilis serum on hospital patients. On the 'Neisser scandal' and its implications for the issue of consent see Chapter 3.
45. Reichsgericht 1905a, pp. 62–65.
46. Cf. Flesch 1905, p. 38.
47. Cf. Moll 1936, pp. 270–271. On Moll's book on medical ethics, see Chapter 4.
48. See Hellwig 1904. In 1903 the I. District Court of Munich had convicted a man to five months imprisonment, because he had had intercourse with a young woman although he knew that he had an uncured venereal disease. Cf. 'Körperverletzung durch geschlechtliche Ansteckung', *Aerztliche Sachverständigen-Zeitung* 9 (1903), p. 324. See also Sauerteig 1999, pp. 361–362.
49. Reichsgericht 1905a, pp. 63–66. The decision was also reported in the *Deutsche Medizinische Wochenschrift*, 29 June 1905, p. 1037, and the *British Medical Journal*, 2 June 1905, p. 1316.
50. [Anon.] 1905, col. 556.
51. In his memoirs Moll indicated that he had been aware of the relevant Supreme Court decision in the divorce case of 1903 and therefore believed that the appeal in the 1905 case might be successful. Cf. Moll 1936, p. 270.
52. Pallaske 1906, cols 295–296; Bendix 1906, pp. 374–376; Schmidt 1907, p. 20; Alsberg 1908, p. 1359; Sauter 1910, p. 266.
53. See Reichsgericht 1894a; Placzek 1894.
54. Cf. Jaeger 1906, cols 801–803; Placzek 1909, pp. 25–26.

55. *Stenographische Berichte über die Verhandlungen des Preußischen Herrenhauses in der Session 1905/06*, Berlin, Julius Sittenfeld, 1906, p. 265 (14. Sitzung am 31. März 1906).
56. Ibid., pp. 265–266.
57. Cf. Preußischer Ehrengerichtshof 1908–34, vol. 1, pp. 93–98 (decision of 27 September 1907).
58. *Stenographische Berichte über die Verhandlungen des Preußischen Herrenhauses in der Session 1905/06*, Berlin, Julius Sittenfeld, 1906, pp. 267–268.
59. See Sauerteig 1999, pp. 166–186.
60. Cf. Vollmann 1916, col. 458. In defence of notification to the advice centres, see Blaschko 1916. In 1925, the Prussian Court of Honour for Doctors confirmed the disciplinary punishment of a doctor who had reported without direct evidence of their infection two women as sources of venereal disease to the public health bureau of a local insurance organisation. See Preußischer Ehrengerichtshof 1908–34, vol. 4, pp. 55–56 (decision of 14 February 1925).
61. Placzek 1893; Placzek 1909.
62. Zschok 1903; Jummel 1903; Schmidt 1907; Seréxhe 1906; Gans 1907; Exner 1909; Weizmann 1909.
63. Jellinek 1906.
64. Sauter 1910.
65. Cf. Reichsgericht 1886, p. 61.
66. Sauter 1910, pp. 311–318.
67. On the failure of the penal law reform see Chapter 3 in connection with the issue of information and consent.
68. Aschaffenburg 1901, p. 475; Krauß 1904/05b, pp. 160–161, 167–170; Placzek 1909, pp. 30–31, 214–215; Exner 1909, p. 41.
69. Aschaffenburg 1901, pp. 475–476; Krauß 1904/05b, pp. 161–165; Placzek 1909, p. 215.
70. On the development of academic psychiatry in Imperial Germany see Engstrom 2003.
71. Aschaffenburg 1901, p. 476; Moll 1902, pp. 111–112; Krauß 1904/05b, pp. 166–167; Placzek 1909, pp. 207–210. See also the legal advice of seeking consent for demonstrating patients and publishing their case histories in Böhme 1900, pp. 751–752.
72. For a discussion of this movement see Schmiedebach 1996.
73. Peiper 1906, p. 65; Placzek 1909, p. 151.
74. Woycke 1988, pp. 71–72; Elkeles 1989, pp. 76–83. See also Chapter 4.
75. Marx 1876a, p. 53.
76. Cf. Schlegtendal 1895, pp. 503–505.
77. Reichsgericht 1914.
78. Moll 1911.
79. Placzek 1893, p. 55.
80. Cf. Placzek 1909, p. 151.
81. Ibid., pp. 151–154.
82. Cf. Moll 1902, pp. 105–106. For Moll's general position on the issue of abortion, see Chapter 4.

83. Placzek 1909, p. 152. According to Seidler 1993 one may estimate that approximately 300,000 to 500,000 illegal abortions were carried out in late nineteenth-century Germany every year. However, there were on average less than 1,000 convictions per year for criminal abortion during the period 1882 to 1912.
84. Putzke 2003, pp. 382–383.
85. See Usborne 1992, pp. 156–201. See also Grossmann 1995; Koch 2004; Usborne 2007.
86. Cf. Lehmann 1928. For a discussion of the path to this law see Sauerteig 1999, pp. 338–343.

Chapter 3

1. See Elkeles 1996a, pp. 133–148; Gradmann 2005, pp. 134–229 and Gradmann 2001. For an overview of innovations in nineteenth-century surgery see Tröhler 1993.
2. Oppenheim 1892, pp. 7–8. See also Maehle 2000, pp. 209–211; Maehle 2003b, pp. 179–180 and Noack 2004, pp. 23–25.
3. See Keßler 1884, pp. 1–8.
4. Reichsgericht 1880; Hälschner 1881, pp. 469–471; Reichsgericht 1882; Liszt 1884, p. 125; Keßler 1884, pp. 76–77; Oppenheim 1892, pp. 9–10.
5. Oppenheim 1892, pp. 12–13, following in this Keßler 1884, pp. 77–78.
6. Oppenheim 1892, pp. 10–12.
7. Binding 1881, pp. 801–802; Liszt 1884, pp. 124–125; Oppenheim 1892, pp. 14–16.
8. Oppenheim 1892, pp. 17–19, 21–23, 35; Oppenheim 1893; Binding 1881, p. 802. On Waller's experimentation on this and other children and the contemporary scientific and moral debate about it see Elkeles 1996a, pp. 46–52.
9. Oppenheim 1892, p. 4.
10. Ibid., pp. 43–63.
11. See Elkeles 1996a, pp. 140–144; Gradmann 2005, pp. 206–211. However, tuberculin treatment of tuberculous prisoners against their will was forbidden in a Circular of the Prussian Minister of the Interior, dated 28 January 1891; see Winau 1996, pp. 19–21.
12. Oppenheim 1892, p. 4.
13. [Stooss] 1892, p. 465.
14. Stooss 1893, pp. 58–59.
15. Ibid., pp. 57, 60.
16. Ibid., pp. 54, 60–61. The view that surgical interventions could not be classified as physical injuries in the legal sense had also been expressed, in 1892, by the Hamburg lawyer Anton Hess; see Noack 2004, p. 25.
17. Oppenheim 1893, pp. 347–348.
18. Cf. ibid., p. 348.
19. Stooss 1894.
20. Nolte 2007.

21. Nasse 1889, pp. 936–947.
22. Sinn 2001, pp. 71–73.
23. For a full discussion of this problem see Endemann 1893. See also Elkeles 1989, pp. 75–76.
24. Reichsgericht 1894b, p. 376. See also Noack 2004, pp. 26–37.
25. Cf. Verdict of the Hamburg District Court of 2 February 1894, printed in Stooss 1898, pp. 108–112. The process of bone tuberculosis was not stopped, however, by this operation, and the foot had to be amputated by another surgeon a few weeks later. Cf. Reichsgericht 1894b, p. 376.
26. Keßler 1884, p. 78.
27. For details see Noack 2004, pp. 30–33.
28. Cf. Reichsgericht 1894b, p. 382.
29. When Mr K. arrived at the hospital, his daughter was already under narcosis in the operating theatre and her leg had been made bloodless (Esmarch's method). Restoring the blood-flow at this point without having removed the purulent bones would have carried the risk of septicaemia (blood poisoning). Cf. Verdict of the Hamburg District Court of 13 December 1894, printed in Stooss 1898, pp. 118–126.
30. Noack 2004, pp. 33–35.
31. Thiersch 1894, col. 479. Emphasis as in the original.
32. [Wallichs] 1894, col. 497.
33. Cf. Angerer 1899, pp. 351, 356. Emphasis as in the original.
34. Scholz 1914, p. 126.
35. Sinn 2001, pp. 120–121, 134–143; Noack 2004, pp. 98–129. See also Prüll and Sinn 2002.
36. Cf. Thiersch 1894, col. 476.
37. Cf. Angerer 1899, p. 353.
38. Cf. König 1895, p. 6.
39. Angerer 1899, p. 355.
40. König 1895, pp. 2, 5, 7–9.
41. See Elkeles 1996b; Wiesemann 1997.
42. Cf. Elkeles 1989 and Sinn 2001, who quote several further examples of German surgeons' attitudes to consent. For nineteenth- and twentieth-century America see the conflicting studies of Pernick 1982 and Katz 1986.
43. Cf. Verdict of High Court (Oberlandesgericht) Dresden of 7 October 1897, printed in Tränkner 1899. See also Ewald 1899 and Noack 2004, pp. 38–39.
44. Cf. *Dresdener Nachrichten*, Nr. 50, 19 February 1899, in Geheimes Staatsarchiv Preussischer Kulturbeseitz (GStAPK) I. HA Rep. 76 VIII B, Nr. 785.
45. Ewald 1899, p. 139; Noack 2004, p. 39.
46. *Berliner Börsenzeitung*, Nr. 43, 1899, and 'Der Fall Ihle', offprint from *Ärztliche Rundschau* (edited by Dr. med. Arno Krüche, Munich), Nr. 8, 1899, pp. 1–8, both in GStAPK I. HA Rep. 76 VIII B, Nr. 785. See also Noack 2004, pp. 40–41.
47. Ewald 1899, pp. 139–140.

48. Stenglein 1899a; Stenglein 1899b.
49. Stooss 1898; Stooss 1899a; Stooss 1899b, 1899c.
50. [Anon.] 1899.
51. Reichsgericht 1905b, pp. 34–35.
52. Boyens 1907; Seelig and Scheele 1907; Reichsgericht 1908; Seelig and Hacke 1911; [Anon.] 1911; Ebermayer 1925, pp. 127–138. See also Sinn 2001, pp. 47–52 and Noack 2004, pp. 54–59.
53. For more detailed discussions of the various legal theories see Bockelmann 1981 and Maehle 2000.
54. See Lilienthal 1899; Bar 1902; Zitelmann 1907.
55. See Heimberger 1899; Schmidt 1900; Stooss 1902; Stooss 1903; Kahl 1909.
56. See for example the theses by Dietrich 1896, Ehlert 1909, Herzberg 1913; and the articles by Finger 1900, Brückmann 1902, Rosenberg 1903, Behr 1903, Brückmann 1904, Katz 1911, Lieske 1912. See also the comprehensive overview and discussion by Spinner 1914, pp. 185–283.
57. For Quidde's biography, see Holl 2007.
58. Cf. Elkeles 1996a, pp. 190–206; see also Elkeles 1985, Elkeles 2001, Elkeles 2004 and Sauerteig 2000, pp. 307–309.
59. Cf. Elkeles 1996a, pp. 206–208, 221–224.
60. Cf. 'Anweisung an die Vorsteher der Kliniken, Polikliniken und sonstigen Krankenanstalten', 29 December 1900, reprinted in Elkeles 1996a, p. 209.
61. This point has been made by Elkeles 1996a, pp. 206, 215–217, as well as by Sauerteig 2000, pp. 330–331. The Prussian Instruction of 1900 was likewise ignored in medical human subject research in the German colonies in Africa; see Eckart 1997, pp. 161–174, 201–208, 336–349; Eckart and Reuland 2006, p. 38. Much of this research, however, would have been regarded as therapeutic and thus not covered by the instruction.
62. Elkeles 2001, pp. 25–27.
63. Dührssen 1903; Brückmann 1904, p. 658; Sinn 2001, p. 78.
64. See Chapter 4.
65. Scholz 1914, pp. 126–127, 129–133. On medical courts of honour see Chapter 1.
66. Scholz 1914, pp. 126, 128, 134.
67. Moll 1902. On Moll and his book on medical ethics, see in more detail Maehle 2001 and Chapter 4.
68. Cf. Moll 1902, pp. 504–505. Moll named the following countries in which problematic human experiments had been carried out: Germany, Austria, Switzerland, France, Italy, England, Russia, Norway, Sweden, Denmark, Rumania, USA, Chile and Egypt.
69. Moll 1902, pp. 564–568.
70. Ibid., pp. 261–263.
71. Like Moll of Jewish descent, Julius Pagel had achieved the status of an unsalaried university lecturer (*Privatdozent*) in 1891, and was appointed as (still unsalaried) extraordinary professor for history of medicine in 1901. He earned his living as a panel and poor law doctor.

The Berlin chair for history of medicine was given in 1901 to Bismarck's personal physician, Ernst Schweninger. For Pagel's role in the development of history of medicine as an academic discipline see Frewer and Roelcke 2001.

72. Pagel 1897, pp. 41–42, 45–46. For further discussion of Pagel's advice booklet, see Chapter 4.
73. Hundeshagen 1905, p. 298.
74. Peiper 1906, pp. 56–57.
75. Joachim and Korn 1911, pp. 88–98.
76. See also Sinn 2001, pp. 101–115, who provides further examples.
77. See Spinner 1914, pp. 228–229; Riha 1995, p. 11; Noack 2004, p. 94.
78. Müller 2004, pp. 159–169; Wetzell 1996. The 'classical' school of penal law, led by Karl Binding and Wilhelm Kahl, saw punishment as retribution, whereas the 'modern' school, led by Franz von Liszt, aimed at crime prevention through improvement of the criminal and preventive detention.
79. Cf. Sinn 2001, p. 136; Noack 2004, pp. 63–64.
80. Zitelmann 1907, pp. 2144–2145; Sinn 2001, p. 137; Noack 2004, p. 63.
81. Zitelmann 1907, pp. 2145–2147; Noack 2004, p. 64.
82. Kahl 1909, pp. 368–371.
83. Cf. Heimberger 1910, p. 34; Ebermayer 1911a, pp. 1128–1129; Lieske 1912, p. 1570.
84. Cf. Noack 2004, pp. 65–67; Sinn 2001, p. 138. On the American situation see Mohr 1996.
85. Heimberger 1910, p. 35; Ebermayer 1911a, p. 1129; Ebermayer 1911b, p. 1752.
86. Cf. Noack 2004, pp. 68–72. See also Sinn 2001, pp. 140–141.
87. Reichsgericht 1912, pp. 432–433.
88. Cf. ibid., pp. 433–434.
89. See Sinn 2001, pp. 144–149.

Chapter 4

1. See Nutton 1993.
2. See Wear, Geyer-Kordesch and French 1993; Baker, Porter and Porter 1993; Baker 1995; Haakonssen 1997; Strätling 1998; McCullough 1998.
3. See Hoffmann 1752; Gregory 1772; Ploucquet 1797; Percival 1803/1975; Hufeland 1806. Hufeland's essay on the relationships of the physician was later incorporated in his handbook of the practice of medicine, Hufeland 1836, pp. 891–912.
4. Ritzmann 1999.
5. For a comparison between English and German medical deontological literature of the early twentieth century see Schomerus 2001. On medical deontological writings in the nineteenth century see Brand 1977 for Germany and Karstens 1984 for France, and for a general overview Engelhardt 1989.

6. For interpretations and discussions of Bourdieu's concept of 'habitus' see Vogt 1997, pp. 116–121; Robbins 2000, pp. 26–29; Ostrow 2000.
7. Marx 1874 and Marx 1876a. See also Marx 1876b; Marx 1877. For a brief, critical biographical sketch of Marx's life and work see Husemann 1932.
8. Cf. Marx 1874, pp. 5–6; Marx 1876a, p. 46.
9. Marx 1874, pp. 6–10.
10. Ibid., pp. 12–23; Marx 1876a, pp. 15–16, 19–22, 28–29, 69. See also Marx [no date], pp. 121–141. On the German vivisection debate of the nineteenth century see Tröhler and Maehle 1990 and Maehle 1996.
11. Cf. Marx 1876a, p. 1.
12. Ibid., pp. 4–6.
13. Ibid., p. 29.
14. Ibid., pp. 46–47.
15. Ibid., pp. 30–33. On medical confidentiality see in detail Chapter 2.
16. Marx 1876a, pp. 33–34.
17. For overviews of the history of medical truth-telling, see Engelhardt 1996 and Nolte 2008. See also Beauchamp 1995 on the American physician Worthington Hooker, who in 1849 argued against deception, because – if discovered – it might destroy the patient's trust in the doctor, but who justified under-disclosure if full disclosure was believed to harm the patient. Nolte 2008 has shown for nineteenth-century Germany how doctors' reluctance to disclose a bad prognosis was in conflict with the intentions of Protestant nurses who wished terminally ill patients to be informed about their imminent end, so that they could be spiritually prepared for death.
18. Marx 1876a, p. 41; Marx 1827, p. 146. See also Cane 1952.
19. See Stolberg 2007a; Benzenhöfer 1999, pp. 71–76; Engelhardt 2000. However, two early nineteenth-century German medical practitioners who advocated active euthanasia with opiates in terminally ill patients have been identified by Stolberg 2008.
20. For the history and debate of such practices see Stolberg 2007b.
21. Marx 1876a, p. 53.
22. Ibid., pp. 56–57, 65.
23. Cf. Husemann 1932, p. 106. Because of his criticisms of modern scientific medicine Marx was apparently unpopular in the Medical Faculty of Göttingen University; cf. ibid.
24. For a detailed discussion of these issues see Huerkamp 1985. See also Herold-Schmidt 1997.
25. Schmidt [no date; *c.* 1884], pp. 8–9.
26. Ibid., pp. 40–42.
27. Ibid., p. 28. On the 1885 decree, issued by the Prussian Minister for Religious, Educational and Medical Affairs Gustav von Goßler, and its context, see Tröhler and Maehle 1990, and Maehle 1996. The decree required anaesthesia for experimental animals (if this did not contravene the purpose of the experiment), preferably experimentation on lower classes of animals, and limitation of animal experiments to serious research and important teaching purposes only.

28. An early key text in this literary tradition was Gabriele de Zerbi's *De cautelis medicorum* (1495); see French 1993 and Linden 1999.
29. Gersuny 1884, pp. 18–22.
30. Ibid., p. 7.
31. Ibid., pp. 28–29.
32. Ibid., pp. 36–39.
33. See Huerkamp 1985, pp. 78–87.
34. Hasse 1886, pp. 6–10, 70–74.
35. Ibid., pp. 1–5; Huerkamp 1985, pp. 61–78; Herold-Schmidt 1997, pp. 65–67.
36. Huerkamp 1985, p. 62.
37. For a brief historical introduction into this issue see Bleker 1996. As the first of the German states, Baden admitted women to regular medical studies at its universities from 1899/1900, but Prussia did so only from 1908.
38. Cf. Hasse 1886, pp. 41–45, 78–79, 83–84.
39. Ibid., p. 64.
40. Ibid., pp. 22–23, 74–75.
41. Ziemssen 1887. For public lectures by university professors reviewing past and present of the medical profession see for example Haeser 1860; Billroth 1891; Vierordt 1893; Bollinger 1908.
42. Ziemssen, 1887, pp. 6, 15–17. Ziemssen's concerns about a potential increase in student numbers turned out to be well founded. After admission of school-leavers from the *Realgymnasien* and *Oberrealschulen* to medical studies, the number of medical students at Prussian universities rose from 1,975 in 1905/06 to 3,867 in 1911/12, the proportion of medical students coming from these two school-types rising in the same period from 10.7 per cent to 21.4 per cent; cf. Huerkamp 1985, p. 87.
43. See Smith 2007, pp. 127–128.
44. Cf. Ziemssen 1887, p. 18.
45. Ibid., pp. 18–19.
46. Ibid., pp. 10–11.
47. Ibid., pp. 12–13.
48. Ibid., pp. 19–22. Ziemssen fully approved of the then existing structures of professional organisation in Bavaria, including the medical district societies, the doctors' chambers and the enlarged Higher Medical Commission. On the roles of these bodies with regard to professional discipline see Chapter 1.
49. Cf. Tannenbaum 1994 on the similar characteristics expected of medical men in nineteenth-century America.
50. Herold-Schmidt 1997, p. 66.
51. Ziemssen 1887, p. 14.
52. Cf. Vierordt 1893, p. 97.
53. Cf. Wolff 1896, p. 7. Emphasis as in the original.
54. Ibid., pp. 12, 37, 46, 59, 61–62, 68, 70, 94, 111, 113, 128.
55. Ibid., pp. 5, 50–53, 57, 62, 154, 156.
56. See ibid., pp. 137–148.

57. Ibid., pp. 77–78, 105–106.
58. See Chapter 2.
59. Seidler 1993, pp. 135, 137–138.
60. Wolff 1896, pp. 75–76, 96, 119–120.
61. Ibid., pp. 129, 131–132.
62. Pagel 1897, pp. 15–21.
63. Ibid., p. 23. In 1901 the 4th edition of Gersuny's booklet was published; see Brand 1977, p. 206.
64. Mettenheimer 1899, pp. 24–26.
65. Ughetti 1899.
66. Scholz 1927. This 5th edition was edited by Erwin Liek.
67. In an article in a popular magazine Dessoir had called for medical ethics teaching in order to compensate for a loss of ethical values in a depersonalised doctor–patient relationship that resulted from the influence of science in medicine and medical specialisation. See Dessoir 1894. For Dessoir's biography see Hermann 1929 and the autobiography Dessoir 1947.
68. Pagel 1951, pp. 208–209.
69. See Pagel 1899, pp. 26–39.
70. Cf. Pagel 1897, pp. 1–5. Pagel had been practising in Berlin since 1876, in his own practice and as municipal poor law physician, and had become *Privatdozent* (unsalaried university lecturer) for history of medicine at Berlin University in 1891.
71. Ibid., p. 33.
72. Cf. ibid., pp. 34–38.
73. Cf. ibid., pp. 42–45. See also Chapter 3.
74. For further references to monographs and articles of this genre during this period see Pagel 1897, pp. 23–26; Brand 1977; Engelhardt 1989.
75. Cf. Ziemssen 1898, p. 69; Ziemssen 1899, p. 5.
76. Cf. Jaksch 1898, p. 76.
77. Ziemssen 1899, pp. 11–38; Styrap 1878. For a modern edition and discussion of de Styrap's code see Bartrip 1995.
78. Ziemssen 1899, p. 10.
79. Ibid., p. 7.
80. Grassmann 1902, col. 406.
81. Peiper 1906. See also Chapter 3.
82. Moll 1902, p. V.
83. Schultz 1986, pp. 97–99; Eben 1998, pp. 124–127.
84. Hahn 1984, pp. 560–561.
85. For discussion of this issue, see below.
86. For Moll's biography see Winkelmann 1996; Cario 1999; Maehle 2007. See also the autobiography Moll 1936.
87. *Der Hypnotismus* (Berlin 1889), also published in English translation: Moll 1891.
88. See Winkelmann 1992.
89. See also Maehle 2001.
90. Moll 1902, p. VI; Moll 1936, pp. 16–17. On Dessoir and the Society for Experimental Psychology see Kurzweg 1976.

91. *Arzt contra Bakteriologe* (Berlin and Vienna 1903). Cf. Engel 1965. On Rosenbach see also Steinacher 1959.
92. Simmel 1958. For a discussion of Simmel's work around 1900 see Frisby 2002.
93. Moll 1902, pp. 7–10.
94. Ibid., p. 9.
95. Ibid., pp. 10–11.
96. Ibid., pp. 11–14, 33–43. See also Moll 1899/1900.
97. Moll 1902, pp. 124, 211–220.
98. Cf. Hufeland's famous dictum that 'to pronounce death is to give death'; Hufeland 1806, p. 17. For Marx see above and Marx 1876a, pp. 33–34.
99. Moll 1902, pp. 120–127.
100. Benzenhöfer 1999, pp. 80–91. For full discussions of the history of racial hygiene and eugenics in Germany, including their role in National Socialism, see for example Weindling 1989 and Schmuhl 1992.
101. Jost 1895.
102. Jost 1895, pp. 26, 39–44, 47. See also Benzenhöfer 1999, pp. 92–95.
103. Moll 1902, pp. 128–129, 235.
104. Benzenhöfer 1999, p. 77.
105. Moll 1902, pp. 127, 129–133, 243, 515, 558. Such experiments had been carried out, for example, in the late 1880s by Kurt Schimmelbusch of the surgical department of the University of Halle. He had transferred staphylococci to dying persons in order to study the cause of boils. Cf. Elkeles 1996a, p. 186.
106. Moll 1902, pp. 249–259.
107. Ibid., pp. 259–260. For the debate on a legal reform of abortion in Imperial and Weimar Germany see Usborne 1992; Grossmann 1995; Putzke 2003; Koch 2004 and Usborne 2007.
108. Jaffé 1902, p. 287.
109. Most doctors advocated the 'freie Arztwahl', as this strengthened their position vis-à-vis the sickness insurance organisations. See Labisch 1997.
110. Henius 1902, p. 529.
111. Cf. Moll 1936, p. 272.
112. Stenglein 1902. For Stenglein's and Moll's views on the issue of patient consent see Chapter 3.
113. See also Schomerus 2001, pp. 96, 148–149.
114. See Cario 1999.
115. Cf. Schomerus 2001, pp. 171–173.
116. Cf. Nassauer 1908, p. 1790.
117. See for example Knauer 1919 (1st edition 1912); Scholz 1914 (4th edition).
118. See Rapmund and Dietrich 1898/99; Joachim and Korn 1911, see also Chapter 3.
119. Spinner 1914.
120. See Hundeshagen 1905; Peiper 1906.

121. Ebermayer 1925; Ebermayer 1930. On the history of medical malpractice and doctors' liability see Riegger 2007.
122. Schweninger 1906.
123. For discussions of this type of literature see for example Engelhardt 1989; Göckenjan 1989; Schmiedebach 1989; Kater 1990; Leven and Prüll 1994; Wiesing 1996; Hau 2001 and Schomerus 2001, pp. 101–105, 158–169.
124. Wiesing 1996, p. 188.
125. On Schweninger's biography see Rothschuh 1984.
126. Rothschuh 1983, p. 143.
127. Schweninger 1906, pp. 28–46, 68–78; Wiesing 1996, pp. 188–194.
128. See Epilogue.

Epilogue

1. For the social psychological, political and cultural implications of the German defeat see Schivelbusch 2003.
2. See for Prussia and Saxony in detail, Rabi 2002.
3. Luther 1975, p. 24; Rabi 2002, pp. 46–47.
4. Knüpling 1965, pp. 15–20.
5. This included Prussia (1887/1899/1904), Oldenburg (1891), Hamburg (1894), Bavaria (1871/1895/1921/1927), Anhalt (1900/1920), Lubeck (1903), Saxony (1896/1904), Schaumburg-Lippe (1905), Baden (1864/1883/1906), Brunswick (1865/1908), Waldeck-Pyrmont (1912), Lippe-Detmold (1921), Hesse (1924), Wurttemberg (1925), Thuringia (1926) – in brackets the years of relevant decrees or legislation. Among these, Hesse and Wurttemberg had no compulsory membership in a medical representative body (medical chamber or medical district society). Bremen and Mecklenburg had no medical chambers yet. Cf. Finkenrath 1928, pp. 46–55.
6. Hochtritt 1969, pp. 40–42; Vogt 1998, pp. 367–368.
7. For a discussion of the genesis of this law see Benzenhöfer 2006. It has been estimated that over 300,000 people were sterilized on the basis of this law between 1934 and 1945. Cf. ibid., pp. 7–8.
8. Cf. Böttger 1935, p. 76.
9. *Reichsärzteordnung*, 13 December 1935, § 13 (3): 'Der Täter ist straffrei, wenn er ein solches Geheimnis zur Erfüllung einer Rechtspflicht oder sittlichen Pflicht oder sonst zu einem nach gesundem Volksempfinden berechtigten Zweck offenbart und wenn das bedrohte Rechtsgut überwiegt.'
10. See Noack 2004, pp. 113–124; Eckart and Reuland 2006; Weindling 2004.
11. Sauerteig 2001, p. 88.
12. See Noack 2004, pp. 130–147; Sinn 2001, pp. 54–58.
13. Reichsgericht 1936; Reichsgericht 1940; Sinn 2001, pp. 60–61; Maehle 2003b, pp. 184–185; Noack 2004, pp. 160–170, 174–178.
14. Honigmann 1924, pp. 257, 273–275; Wiesing 1996, pp. 194–199.

15. Cf. Wiesing 1996, p. 188. Liek ran a private surgical and gynaecological practice in Danzig, which he gave up in 1931 because of poor health, subsequently living as a medical writer in Berlin until his death in 1935 at the age of 56.
16. Cf. Liek in Scholz 1927, p. XI.
17. Nassauer 1911.
18. Nassauer 1925, pp. 11–15.
19. Cf. Haeberlin 1925 (1st edition 1919), pp. 92–95; Krecke 1934 (English translation of the 2nd German edition of 1932), pp. 33–46.
20. Cf. Nassauer 1925, p. 41.
21. See Schmiedebach 1989, pp. 46–47; Kater 1990, pp. 445–446.
22. Binding and Hoche 1920. For a discussion of this book and contemporary controversial responses, see for example Benzenhöfer 1999, pp. 100–108.
23. Schmiedebach 2001.
24. Liek 1926, pp. 46–47, 131.
25. Ibid., pp. 61–62, 131.
26. In later editions of his book, from the 4th edition in 1927, Liek included however a separate chapter on racial hygiene. Cf. Schomerus 2001, pp. 103, 160.
27. See Liek 1926.
28. Cf. ibid., pp. 37–40, 73–74, 97, 111.
29. Ibid., pp. 118–123. See on this aspect also Hau 2001.
30. Liek 1926, p. 136.
31. On the influence of Liek's thought on Nazi medicine see Kater 1989, pp. 227–230, 235–236.
32. For full discussions of the German medical profession's path into Nazism and important role in the Third Reich see for example Thomsen 1996 and Kater 1989.
33. Kater 1985. McClelland 1997 has argued against this interpretation that the notion of 'interrupted' professionalisation due to the economic crises of the Weimar Republic, rather than a 'failure' of professionalisation, may better help to explain German doctors' path to Nazism. For further discussion and analysis of this path see Kater 1989; Thomsen 1996. On the *Sonderweg* discussion in general, see for example Kocka 1988.
34. See, for example, Lifton 1986; Proctor 1988; Annas and Grodin 1992; Bleker and Jachertz 1993; Burleigh 1994; Bock 1997; Weindling 2004.
35. See Smith 1994.
36. Styrap 1878 (reprint 1995), p. 154; Bartrip 1995.
37. Styrap 1878, p. 155. See also Morrice 2002a; Binder 2000.
38. See Morrice 2002b; Ferguson 2005; Pranghofer and Maehle 2006.
39. [Anon.] 1896a; [Anon.] 1896b; Moscucci 1993, pp. 162–164.

Bibliography

Archival sources

Bayerisches Hauptstaatsarchiv (BayHStA), Munich
MInn, Nr. 60399, 61413, 61414, 61415, 61416, 61420, 61421, 61427, 61428; MJu Nr. 15367, 15393.
Geheimes Staatsarchiv Preussischer Kulturbesitz (GStAPK), Berlin
I. HA Rep. 76 VIII B, Nr. 591, 592, 594, 770, 771, 782, 783, 784, 785, 793, 830; I. HA Rep. 84a, Nr. 101.
Sächsisches Hauptstaatsarchiv (SächsHStA), Dresden
Min. des Innern, Nr. 15146, 15147, 15199.
Staatsarchiv München (StAM), Munich
RA Nr. 57093, 57094, 57095.

Primary literature

Alsberg, M. (1908) Das ärztliche Berufsgeheimnis, *Deutsche Medizinische Wochenschrift* 34, pp. 1356–1359.

Altmann, F. (1900) *Aerztliche Ehrengerichte und ärztliche Standesorganisation in Preußen. Das Preußische Gesetz betreffend die ärztlichen Ehrengerichte, das Umlagerecht und die Kassen der Aerztekammer vom 25. November 1899 zum praktischen Handgebrauch erläutert*, Berlin, H. W. Müller.

Angerer, O. von (1899) Die strafrechtliche Verantwortlichkeit des Arztes, *Münchener Medicinische Wochenschrift* 46, pp. 351–356.

[Anon.] (1896a) Beatty Versus Cullingworth, *Lancet*, Nov. 21, pp. 1473–1474.

[Anon.] (1896b) Beatty v. Cullingworth, *British Medical Journal*, Nov. 21, pp. 1525–1526.

[Anon.] (1899) Tagesgeschichtliche Notizen, *Münchener Medicinische Wochenschrift* 46, pp. 203–204, 371–372.

[Anon.] (1903a) Bilz und seine '3 Aerzte' in ehrengerichtlicher Betrachtung, *Ärztliches Vereinsblatt für Deutschland* 30, cols 190–192.

[Anon.] (1903b) The Medical Man and the Law, *Lancet*, Oct. 3, p. 984.

[Anon.] (1905) 'Befugte' Offenbarung eines Privatgeheimnisses seitens eines Arztes (§ 300 Str.-G.B.), *Ärztliches Vereinsblatt für Deutschland* 32, cols 556–558.

[Anon.] (1911) Gerichtliche Entscheidungen. Die Haftung des Arztes. Urteil des Reichsgerichts vom 30. Juni 1911, *Münchener Medizinische Wochenschrift* 58, p. 1943.

154   *Bibliography*

Ärztetag (1881) Verhandlungen des IX. Deutschen Aerztetages zu Cassel am 1. und 2. Juli 1881. (Officielles Protokoll.), *Ärztliches Vereinsblatt für Deutschland* 8, cols 193–207.

Ärztetag (1882) Verhandlungen des X. Deutschen Aerztetages zu Nürnberg am 30. Juni und 1. Juli 1882. (Officielles Protokoll.), *Ärztliches Vereinsblatt für Deutschland* 9, cols 209–225.

Ärztetag (1889) Verhandlungen des XVII. Deutschen Aerztetages zu Braunschweig am 24. und 25. Juni 1889 (Officielles Protokoll.), *Ärztliches Vereinsblatt für Deutschland* 16, cols 273–291, 321–340.

Ärztlicher Bezirksverein München (1875) *Der Aerztliche Stand und das Publikum. Eine Darlegung der beiderseitigen und gegenseitigen Pflichten*, Munich, J. A. Finsterlin.

Aschaffenburg, G. (1901) Berufsgeheimnis (§ 300 St. G. B.) und Psychiatrie, *Aerztliche Sachverständigen-Zeitung* 7, pp. 473–477.

Auerbach, E. (1902) Das Zeugnisverweigerungsrecht weiblicher Aerzte in Strafsachen, *Juristische Wochenschrift* 31, pp. 381–383.

Augustin, F. L. (1838) *Die Königlich preußische Medicinalverfassung oder vollständige Darstellung aller, das Medicinalwesen und die medicinische Polizei in den Königlich Preußischen Staaten betreffenden Gesetze, Verordnungen und Einrichtungen*, vol. 6: *Medicinalverordnungen 1833 bis 1837*, Potsdam.

Ausschuss der preussischen Ärztekammern (1892) Sitzung des Ausschusses der preussischen Aerztekammern am 25. Oktober Morgens 9 Uhr im Provinzialständehause in Berlin, *Ärztliches Vereinsblatt für Deutschland* 19, cols 414–420.

Bar, L. von (1902) Zur Frage der strafrechtlichen Verantwortlichkeit des Arztes, *Der Gerichtssaal* 60, pp. 81–112.

Behr, A. (1903) Medicin und Strafrecht, *Der Gerichtssaal* 62, pp. 400–424.

Bendix, L. (1906) Zur Verschwiegenheitspflicht der Ärzte, *Zeitschrift für Bekämpfung der Geschlechtskrankheiten* 5, pp. 372–376.

Bernstein, M. (1905) Ärztliches Berufsgeheimnis und Geschlechtskrankheiten, *Zeitschrift für Bekämpfung der Geschlechtskrankheiten* 4, pp. 29–31.

Biberfeld, Dr. jur., (1902) Die Schweigepflicht des Arztes, *Zeitschrift für Medizinal-Beamte* 15, pp. 648–650.

Billroth, T. (1891) Arzt, Staat und Publicum, *Wiener Klinische Wochenschrift* 4, pp. 922–926.

Binding, K. (1881) *Handbuch des Strafrechts*, Erster Band, Leipzig, Duncker & Humblot.

Binding, K. (1890) *Die Ehre im Rechtssinn und ihre Verletzbarkeit*, Leipzig, Alexander Edelmann.

Binding, K. and Hoche, A. (1920) *Die Freigabe der Vernichtung lebensunwerten Lebens. Ihr Maß und ihre Form*, Leipzig, Felix Meiner.

Blaschko, A. (1916) Beratungsstellen für Geschlechtskranke, ärztliche Schweigepflicht und Anzeigepflicht, *Ärztliches Vereinsblatt für Deutschland* 43, cols 507–510.

Böhme (Staatsanwalt Dr) (1900) Strafbarkeit der Verletzung des ärztlichen Berufsgeheimnisses, *Allgemeine Zeitschrift für Psychiatrie und psychisch-gerichtliche Medicin* 57, pp. 743–761.

Bollinger, O. von (1908) *Wandlungen der Medizin und des Ärztestandes in den letzten 50 Jahren*, Munich, E. Wolf & Sohn.

Böttger, E. (1935) Die Grenzen der Schweigepflicht des Arztes nach § 7 des Gesetzes zur Verhütung erbkranken Nachwuchses vom 14.7.1933 (RGBl. I, 529), *Ärzteblatt für Berlin* 40, p. 76.

Boyens (Rechtsanwalt beim Reichsgericht) (1907) Reichsgericht. 1. Zivilsachen. Schadensersatzanspruch gegen einen Arzt wegen einer mißlungenen Operation (Urteil v. 21. Mai [Juni] 1907), *Spruch-Beilage zur Deutschen Juristen-Zeitung* 12, col. 1025.

Brauser (, A. G.) (1899) Aus Bayern, *Ärztliches Vereinsblatt für Deutschland* 26, cols 48–51.

Brückmann, A. (1902) Zur Frage der strafrechtlichen Verantwortlichkeit des Arztes für operative Eingriffe, *Deutsche Medicinische Wochenschrift* 28, pp. 328–329, 345–347.

Brückmann, A. (1904) Neue Versuche zum Problem der strafrechtlichen Verantwortlichkeit des Arztes für operative Eingriffe. Negatives und Positives, *Zeitschrift für die gesamte Strafrechtswissenschaft* 24, pp. 657–714.

Chotzen, M. and Simonson (Oberlandesgerichtsrat) (1903/04) Meldepflicht und Verschwiegenheits-Verpflichtung des Arztes bei Geschlechtskrankheiten, *Zeitschrift für Bekämpfung der Geschlechtskrankheiten* 2, pp. 433–474.

Cnyrim, V. (1892) Entgegnung, *Ärztliches Vereinsblatt für Deutschland* 19, cols 129–132.

Coblenzer ärztlicher Bezirksverein (1880) Die Stellung der Aerzte in der Gewerbeordnung, *Ärztliches Vereinsblatt für Deutschland* 7, cols 129–134.

Dessoir, M. (1894) Der Beruf des Arztes, *Westermanns Illustrierte Deutsche Monatshefte* 77, pp. 375–382.

Dessoir, M. (1947) *Buch der Erinnerung*, 2nd edn, Stuttgart, Ferdinand Enke.

Dietrich, H. (1896) *Die Straflosigkeit ärztlicher Eingriffe*, Jur. Diss. Marburg, Fulda, Fuldaer Aktiendruckerei.

Dittenberger, H. (1906) Zum § 300 des Reichsstrafgesetzbuches, *Monatsschrift für Kriminalpsychologie und Strafrechtsreform* 2, pp. 54–58.

Doerfler, H. (1908) Verbesserung oder Neuorganisation unserer bayerischen ärztlichen Ehrengerichte?, *Münchener Medizinische Wochenschrift* 54, pp. 1997–1998.

Dührssen, A. (1903) Strafgesetzbuch und ärztliche Operationen, *Berliner Aerzte-Correspondenz* 8, pp. 13–14.

Ebermayer, L. (1910) Die Unruhen in Berlin-Moabit und das Zeugnisverweigerungsrecht der Aerzte, *Ärztliches Vereinsblatt für Deutschland* 37, cols 828–829.

Ebermayer, L. (1911a) Die Stellung des Arztes im Vorentwurf zu einem deutschen Strafgesetzbuche, *Deutsche Medizinische Wochenschrift* 37, pp. 1128–1131.

Ebermayer, L. (1911b) Rechtsfragen aus der ärztlichen Praxis, *Deutsche Medizinische Wochenschrift* 37, pp. 1752–1753.

Ebermayer, L. (1925) *Arzt und Patient in der Rechtsprechung*, 3rd edn, Berlin, Rudolf Mosse.

Ebermayer, L. (1930) *Der Arzt im Recht. Rechtliches Handbuch für Ärzte*, Leipzig, Georg Thieme.

Ehlert, K. (1909) *Worin liegt der Grund für die Rechtmässigkeit eines ärztlichen Eingriffes, der ohne Einwilligung des Behandelten oder seines Vertreters vorgenommen wird?* Law thesis, Ruprecht-Karls-Universität Heidelberg, Berlin, W. Pilz.

Endemann, F. (1893) *Die Rechtswirkungen der Ablehnung einer Operation seitens des körperlich Verletzten. Ein Beitrag zur Lehre von der civilrechtlichen Haftung aus Körperverletzungen und zur Auslegung der Reichsversicherungsgesetze*, Berlin, Carl Heymanns.

Eulenburg, A. (1896) Zur Stellung der Aerzte vor und nach der Gewerbeordnung von 1869, *Deutsche Medicinische Wochenschrift* 22, pp. 714–716.

Ewald, C. A. (1899) Ueber den Fall I. in Dresden, *Berliner Klinische Wochenschrift* 36, pp. 139–140.

Exner, B. (1909) *Das Berufsgeheimnis des Arztes gemäß § 300 des Str. G. B.*, Jur. Diss. Heidelberg, Heidelberger Verlagsanstalt und Druckerei.

Finger, A. (1900) Chirurgische Operation und ärztliche Behandlung, *Zeitschrift für die gesamte Strafrechtswissenschaft* 20, pp. 12–32.

Finkenrath, K. (1928) *Die Organisation der deutschen Ärzteschaft. Eine Einführung in die Geschichte und den gegenwärtigen Aufbau des wissenschaftlichen, standes- und wirtschaftspolitischen ärztlichen Vereinslebens*, Berlin, Fischers Medizinische Buchhandlung H. Kornfeld.

Flesch, M. (1905) Das ärztliche Berufsgeheimnis und die Bekämpfung der Geschlechtskrankheiten, *Zeitschrift für Bekämpfung der Geschlechtskrankheiten* 4, pp. 32–51.

Fromme (Landgerichtsdirektor, Magdeburg) (1902) *Die rechtliche Stellung des Arztes und seine Pflicht zur Verschwiegenheit im Beruf* (Berliner Klinik, Heft 165), Berlin, Fischer.

Gabriel, A. (1919) *Die staatliche Organisation des Deutschen Aerztestandes*, Berlin, Adler-Verlag.

Gaertner, M. (1896) *Staatliche Ehrengerichte für die Aerzte*, Breslau, Preuss & Jünger.

Gans, R. O. (1907) *Das ärztliche Berufsgeheimnis des § 300 RStrGB*, Jur. Diss. Heidelberg, Borna-Leipzig, Buchdruckerei Robert Noske.

Gersuny, R. (1884) *Arzt und Patient. Winke für Beide*, Stuttgart, Ferdinand Enke.

Graf, E. (1880) Aphorismen zur Medicinalreform, *Ärztliches Vereinsblatt für Deutschland* 7, cols 47–51.

Graf, E. (1890) *Das ärztliche Vereinswesen in Deutschland und der Deutsche Ärztevereinsbund. Festschrift dem 10. Internationalen Medizinischen Kongress gewidmet vom Geschäftsausschuss des Deutschen Ärztevereinsbundes und im Auftrage desselben verfasst von dem Vorsitzenden*, Leipzig, F. C. W. Vogel.

Grassmann (Munich) (1902) Ist ärztliche Ethik lehrbar?, *Ärztliches Vereinsblatt für Deutschland* 31, cols 406–408.

Gregory, J. (1772) *Lectures on the Duties and Qualifications of a Physician*, London, W. Strahan and T. Cadell.

Gross, H. (1903) Zur Frage des Berufsgeheimnisses, *Archiv für Kriminal-Anthropologie und Kriminalistik* 13, pp. 241–247.

Guttstadt, A. (1890) *Deutschlands Gesundheitswesen. Organisation und Gesetzgebung des Deutschen Reichs und seiner Einzelstaaten*, Erster Theil, Leipzig, Georg Thieme.

H., Dr (1903) Ueber Verschwiegenheitspflicht und Zeugnisverweigerungsrecht des Arztes vor Gericht, *Ärztliches Vereinsblatt für Deutschland* 30, cols 248–251.

Haeberlin, C. (1925) *Vom Beruf des Arztes*, 2nd edn, Munich, Verlag der Ärztlichen Rundschau Otto Gmelin.

Haeser, H. (1860) *Ueber das Sittliche im Berufe des Arztes*, Greifswald, F. W. Kunike.

Hälschner, H. (1881) *Das gemeine deutsche Strafrecht systematisch dargestellt*, Erster Band, Bonn, Adolph Marcus.

Hasse, C. (1886) *Aus dem Ärztlichen Leben. Ratgeber für angehende und junge Ärzte*, Berlin and Neuwied, Heuser's Verlag.

Heger, A. (1940) *Berufsgeheimnis und Abtreibung*, Med. Diss. Würzburg, Richard Mayr.

Heimberger, J. (1899) *Strafrecht und Medizin*, Munich, C. H. Becksche Verlagsbuchhandlung.

Heimberger, J. (1910) Der Vorentwurf zu einem Deutschen Strafgesetzbuch in seiner Bedeutung für den Arzt, *Deutsche Medizinische Wochenschrift* 36, pp. 33–35, 78–81.

Heinze, O. (1899) Die Neu-Organisation des ärztlichen Standes in Bayern, *Ärztliches Vereinsblatt für Deutschland* 26, cols 421–430.

Heinze, O. (1901) Die Bilz'sche 'Naturheilanstalt' und deren 'approbirte Aerzte', *Ärztliches Vereinsblatt für Deutschland* 28, cols 549–551.

Heinze, O. (1918) *Der Deutsche Aerztevereinsbund und die ärztlichen Standesvertretungen in Deutschland von 1890 bis 1912*, Leipzig, Selbstverlag des Verfassers.

Hellwig, A. (1904) Die civilrechtliche Bedeutung der Geschlechtskrankheiten, *Zeitschrift für Bekämpfung der Geschlechtskrankheiten* 1, pp. 26–63.

Hellwig, A. (1910) Beschlagnahme ärztlicher Krankenjournale nach geltendem und künftigem Recht, *Deutsche Medizinische Wochenschrift* 36, pp. 2152–2153.

Hellwig, A. (1936) Die Bedeutung der Reichsärzteordnung für die Rechtspflege, *Deutsche Justiz* 98, pp. 370–373.

Henius, L. (1902) Rezension: Albert Moll, Ärztliche Ethik, *Deutsche Medicinische Wochenschrift* 28, pp. 528–529.

Herzberg, C. (1913) *Die Grenzen der Straflosigkeit ärztlicher Eingriffe*, Law thesis, Ruprecht-Karls-Universität Heidelberg.

Hoffmann, A. (1880) Entwurf zu einer Aerzteordnung für das deutsche Reich, *Ärztliches Vereinsblatt für Deutschland* 7, cols 233–238.

Hoffmann, F. (1752) *Politischer Medicus, oder Klugheits-Regeln, Nach welchen ein junger Medicus seine Studia und Lebensart einrichten soll, wenn er sich will berühmt machen, auch geschwinde eine glückliche Praxin zu erlangen und zu erhalten begehret*, transl. into German by J. M. Auerbach, Leipzig, Friedrich Lankischens Erben.

Honigmann, G. (1924) *Das Wesen der Heilkunde. Historisch-Genetische Einführung in die Medizin. Für Studierende und Ärzte*, Leipzig, Felix Meiner.

Hufeland, C. W. (1806) Die Verhältnisse des Arztes, *Journal der practischen Heilkunde* 23, 3. Stück, pp. 5–36.

Hufeland, C. W. (1836) *Enchiridion medicum oder Anleitung zur medizinischen Praxis. Vermächtnis einer funfzigjährigen Erfahrung*, 2nd edn, Berlin, Jonas Verlagsbuchhandlung.

Hundeshagen, K. (1905) *Einführung in die ärztliche Praxis vom Gesichtspunkte der praktischen Interessen des Ärztestandes unter eingehender Berücksichtigung der Versicherungsgesetze und allgemeinen Gesetzgebung. Für Studierende der Medizin und junge Ärzte*, Stuttgart, Ferdinand Enke.

Husemann, T. (1932) Marx, Karl Friedrich Heinrich, in A. Hirsch (ed.), *Biographisches Lexikon der hervorragenden Ärzte aller Zeiten und Völker*, 2nd edn, Berlin and Vienna, Urban & Schwarzenberg, vol. 4, pp. 105–107.

Jaeger (Staatsanwaltschaftsrat, Metz) (1906) Das Berufsgeheimnis der Aerzte und Anwälte, *Deutsche Juristen-Zeitung* 11, cols 800–805.

Jaffé, K. (1902) Rezension: Albert Moll, Ärztliche Ethik, *Münchener Medicinische Wochenschrift* 49, p. 287.

Jaksch, R. von (1898) Ueber den medicinisch-klinischen Unterricht. Schlusswort, in *Verhandlungen des Congresses für Innere Medicin* 16, pp. 74–76.

Jellinek, W. (1906) Der Umfang der Verschwiegenheitspflicht des Arztes und des Anwalts, *Monatsschrift für Kriminalpsychologie und Strafrechtsreform* 3, pp. 656–693.

Joachim, H. (1902) Die jüngsten Entscheidungen des Ehrengerichtshofes. Nach einem im ärztlichen Standesverein der Friedrichstadt erstatteten Referat, *Berliner Ärzte-Correspondenz* 7, pp. 225–226.

Joachim, H. (1903) Das Urteil des Ehrengerichtshofes betr. 'die Postkarte an den Oberpräsidenten', *Berliner Ärzte-Correspondenz* 8, p. 19.

Joachim, H. [1913] *Denkschrift über die Tätigkeit der Ärztekammer für die Provinz Brandenburg und den Stadtkreis Berlin in den ersten 25 Jahren ihres Bestehens im Auftrage des Vorstandes der Ärztekammer verfasst*, Berlin, A. Haussmann.

Joachim, H. and Korn, A. (1911) *Deutsches Ärzterecht mit Einschluß der landesgesetzlichen Bestimmungen. Handbuch für Ärzte und Juristen*, Erster Band, Berlin, Franz Vahlen.

Jost, A. (1895) *Das Recht auf den Tod. Sociale Studie*, Göttingen, Dieterich'sche Verlagsbuchhandlung.

Jummel, F. O. (1903) *Der § 300 Str.G. B., ein Versuch seiner Auslegung*, Jur. Diss. Leipzig, Borna-Leipzig, Buchdruckerei Robert Noske.

Kade, C. (1906) *Die Ehrengerichtsbarkeit der Aerzte in Preussen. Eine Bearbeitung des Ehrengerichtsgesetzes und der veröffentlichten Entscheidungen des ärztlichen Ehrengerichtshofes*, Berlin, August Hirschwald.

Kaestner, P. (1911) Der ärztliche Ehrengerichtshof und das erste Jahrzehnt seiner Rechtsprechung, *Ministerial-Blatt für Medizinal- und medizinische Unterrichtsangelegenheiten* 11, pp. 352–354.

Kahl, W. (1909) Der Arzt im Strafrecht, *Zeitschrift für die gesamte Strafrechtswissenschaft* 29, pp. 351–371.

Katz, Dr. jur., (1911) Der Operationsvertrag, *Berliner Klinische Wochenschrift* 48, pp. 768–769.

Keßler, R. (1884) *Die Einwilligung des Verletzten in ihrer strafrechtlichen Bedeutung*, Berlin and Leipzig, J. Guttentag.

Knauer, G. (1919) *Winke für den ärztlichen Weg aus zwanzigjähriger Erfahrung*, 2nd edn, Wiesbaden, J. F. Bergmann.

Kohler, J. (1904) Stellung der Rechtsordnung zur Gefahr der Geschlechtskrankheiten, *Zeitschrift für Bekämpfung der Geschlechtskrankheiten* 2, pp. 19–30.

König, F. (1895) Der Arzt und der Kranke. Mit besonderer Berücksichtigung des Krankenhausarztes, *Zeitschrift für sociale Medicin* 1, pp. 1–11.

Köppen, A. (1901) Ueber Standesbewusstsein und Standesehre, *Ärztliches Vereinsblatt für Deutschland* 28, cols 148–150.

Krahmer (Geh. Medizinalrat und Stadtphysikus in Halle) (1875) Der Zeugnisszwang der Aerzte, *Berliner Klinische Wochenschrift* 12, pp. 365–366, 378.

Krauß, R. (1904/05a) Das Berufsgeheimnis und Zeugnisverweigerungsrecht des Arztes und Rechtsanwalts, *Monatschrift für Kriminalpsychologie und Strafrechtsreform* 1, pp. 128–130.

Krauß, R. (1904/05b) Das Berufsgeheimnis des Psychiaters, *Monatschrift für Kriminalpsychologie und Strafrechtsreform* 1, pp. 151–171.

Krecke, A. (1934) *The Doctor and His Patients*, London, Kegan Paul, Trench, Trubner & Co.

Lehmann, R. (1928) Die Schweigepflicht des Arztes und das Gesetz zur Bekämpfung der Geschlechtskrankheiten, *Die Medizinische Welt*, no. 20, pp. 776–778.

Liebmann, J. (1886) *Die Pflicht des Arztes zur Bewahrung anvertrauter Geheimnisse*, Frankfurt/Main, Joseph Baer & Co.

Liek, E. (1926) *Der Arzt und seine Sendung. Gedanken eines Ketzers*, 2nd edn, Munich, J. F. Lehmann.

Lieske, H. (1912) Der ärztliche Eingriff im Spiegel des Rechts, *Berliner Klinische Wochenschrift* 49, pp. 1570–1574.

Lilienthal, K. von (1899) Die pflichtmässige ärztliche Handlung und das Strafrecht, in Jurististische Fakultät der Universität Heidelberg (ed.), *Festgabe zur Feier des fünfzigsten Jahrestages der Doktor-Promotion des Geheimen Rates Professor Dr. Ernst Immanuel Bekker*, Berlin, O. Haering, pp. 1–57.

Liszt, F. von (1884) *Lehrbuch des Deutschen Strafrechts*, Zweite durchaus umgearbeitete Auflage, Berlin and Leipzig, J. Guttentag.

Litten, Dr. jur., (1904/05) Zur Frage des ärztlichen Berufsgeheimnisses, *Monatsschrift für Kriminalpsychologie und Strafrechtsreform* 1, pp. 55–60.

Marcus, Dr (1884) Wegen Verweigerung des Zeugnisses, *Ärztliches Vereinsblatt für Deutschland* 11, cols 94–96.

Marx, K. F. H. (1827) Ueber Euthanasie, *Litterarische Annalen der gesammten Heilkunde* 7, pp. 129–151.

Marx, K. F. H. (1874) *Gegen nicht zu billigende Angewöhnungen und Richtungen der jetzigen Aerzte*, Göttingen, Dieterich.

Marx, K. F. H. (1876a) *Ärztlicher Katechismus. Über die Anforderungen an die Ärzte*, Stuttgart, Ferdinand Enke.

Marx, K. F. H. (1876b) *Aussprüche eines Heilkundigen über Vergangenes, Gegenwärtiges und Künftiges*, Göttingen, Dieterich.

Marx, K. F. H. (1877) *Aphorismen über Thun und Lassen der Aerzte und des Publikums*, Stuttgart, Ferdinand Enke.

Marx, K. F. H. [no date] *Bemerkungen über inneres und äusseres Leben als Winke zur Einsicht und Vorsicht. Nebst einem Gespräche über die Stellung der Aerzte in der Gegenwart und Zukunft*, Göttingen, Dieterich.

Mayer, W. (1908) Zur Verbesserung unserer bayerischen ärztlichen Ehrengerichte, *Münchener Medizinische Wochenschrift* 54, pp. 1788–1790.

Medizinal-Abteilung des Ministeriums (1905–15) *Das Gesundheitswesen des Preußischen Staates* (1903–1913), Berlin, Richard Schoetz.

Mestrum, X. (1881) Keine 'Ehrengerichte', sondern 'Ehrenräthe'!, *Ärztliches Vereinsblatt für Deutschland* 8, cols 170–178.

Mettenheimer, C. von (1899) *Viaticum, Erfahrungen und Rathschläge eines alten Arztes, seinem Sohn beim Eintritt in die Praxis mitgegeben*, Berlin, August Hirschwald.

Mittermaier, W. (1936) Das ärztliche Berufsgeheimnis nach der Reichsärzteordnung, *Monatsschrift für Kriminalpsychologie und Strafrechtsreform* 27, pp. 153–154.

Moll, A. (1891) *Hypnotism*, London, Walter Scott.

Moll, A. (1899/1900) Aufsätze zu einer medicinischen Ethik, *Deutsche Medicinische Wochenschrift* 25, pp. 462–464, and 26, pp. 73–75.

Moll, A. (1902) *Ärztliche Ethik. Die Pflichten des Arztes in allen Beziehungen seiner Thätigkeit*, Stuttgart, Ferdinand Enke.

Moll, A. (1911) Neuere Fragen zum ärztlichen Berufsgeheimnis, *Berliner Aerzte-Correspondenz* 16, pp. 1–4.

Moll, A. (1936) *Ein Leben als Arzt der Seele. Erinnerungen*, Dresden, Carl Reissner.

Mulert, Dr (1904) Bemerkungen zu den Entscheidungen des preussischen ärztlichen Ehrengerichtshofes, *Ärztliches Vereinsblatt für Deutschland* 31, cols 337–340.

Nassauer, M. (1908) Ethische Fragen für den ärztlichen Stand. Zeitgemässe Revision der Etikettenfragen, *Münchener Medizinische Wochenschrift* 54, pp. 1790–1792.

Nassauer, M. (1911) *Sterben … ich bitte darum!*, Munich, Verlag Otto Gmelin.

Nassauer, M. (1925) *Die Doktorschule. Das ist Der Arzt der großen und der kleinen Welt und Die hohe Schule für Aerzte und Kranke in 4. Auflage*, Munich, Verlag der Aerztlichen Rundschau Otto Gmelin.

Nasse, D. (1889) Die Sarkome der langen Extremitätenknochen, *Archiv für Klinische Chirurgie* 39, pp. 886–950.

Neisser, A. (1905) Abänderung des § 300 des Reichs-Strafgesetzbuches und ärztliches Anzeigerecht in ihrer Bedeutung für die Bekämpfung der

Geschlechtskrankheiten, *Zeitschrift für Bekämpfung der Geschlechtskrankheiten* 4, pp. 1–28.

Olshausen, J. (1892) *Kommentar zum Strafgesetzbuch für das Deutsche Reich*, 4th edn, vol. 2, Berlin, Franz Vahlen.

Oppenheim, L. (1892) *Das ärztliche Recht zu körperlichen Eingriffen an Kranken und Gesunden*, Basel, Benno Schwabe.

Oppenheim, L. (1893) Die rechtliche Beurteilung der ärztlichen Eingriffe, *Zeitschrift für Schweizer Strafrecht* 6, pp. 332–352.

Ottmer, F. (1902) *Das Schweigen. Erzählung*, Berlin, Concordia Deutsche Verlags-Anstalt.

Pagel, J. (1897) *Medicinische Deontologie. Ein kleiner Katechismus für angehende Praktiker*, Berlin, Oscar Coblentz.

Pagel, J. (1899) *Einführung in das Studium der Medicin (Medicinische Encyklopädie und Methodologie). Vorlesungen gehalten an der Universität zu Berlin*, Berlin and Vienna, Urban & Schwarzenberg.

Pallaske (Justizrat) (1906) Die Schweigepflicht des Arztes, *Deutsche Juristen-Zeitung* 11, cols 293–297.

Peiper, E. (1906) *Der Arzt. Einführung in die ärztlichen Berufs- und Standesfragen. In XVI Vorlesungen*, Wiesbaden, J. F. Bergmann.

Percival, T. (1803/1975) *Medical Ethics; or, a Code of Institutes and Precepts, Adapted to the Professional Conduct of Physicians and Surgeons* (Manchester 1803), in C. D. Leake (ed.), Huntington, New York, Robert E. Krieger.

Pistor, M. (1890) *Deutsches Gesundheitswesen. Festschrift zum X. internationalen medizinischen Kongress Berlin 1890*, Berlin, Julius Springer.

Placzek, S. (1893) *Das Berufsgeheimnis des Arztes*, Leipzig, Georg Thieme.

Placzek, S. (1894) Der Fall Brinkmann und seine Consequenzen für das ärztliche Handeln, *Ärztliches Vereinsblatt für Deutschland* 21, cols 479–481.

Placzek, S. (1909) *Das Berufsgeheimnis des Arztes*, 3rd edn, Leipzig, Georg Thieme.

Ploucquet, W. G. (1797) *Der Arzt, oder über die Ausbildung, die Studien, Pflichten, Sitten und die Klugheit des Arztes*, Tübingen, J. G. Cotta'sche Buchhandlung.

Preußischer Ehrengerichtshof für Ärzte (1901) Beschlüsse vom 18. Mai und 1. Juli 1901: 'Ein Arzt, welcher fortgesetzt oder in marktschreierischer Weise seine Berufsthätigkeit in der Presse annonciert, macht sich einer Verfehlung gegen die ärztliche Standesehre schuldig', *Ministerial-Blatt für Medizinal- und medizinische Unterrichtsangelegenheiten* 1, p. 249.

Preußischer Ehrengerichtshof für Ärzte (1903a) Urteile vom 2. Dezember 1902: '1. Das häufige Annonzieren enthält eine vom Standpunkte der ärztlichen Standesehre zu starke Betonung und Hervorhebung des gewerblichen Moments, welche geeignet ist, das Ansehen des Standes in den Augen der Bevölkerung herabzusetzen. 2. Die Untersagung des häufigen und reklamehaften Annonzierens steht weder mit den Bestimmungen der Reichs-Gewerbeordnung noch mit dem Gesetze zur Bekämpfung unlauteren Wettbewerbs im Widerspruch.', *Ministerial-Blatt für Medizinal- und medizinische Unterrichtsangelegenheiten* 3, p. 218.

Preußischer Ehrengerichtshof für Ärzte (1903b) Beschluß vom 1. Dezember 1902: 'Die Verweigerung ärztlicher Hilfeleistung in Fällen dringender Lebensgefahr, mag diese durch eine plötzliche schwere Erkrankung oder durch die plötzliche Verschlimmerung einer bereits bestehenden Krankheit herbeigeführt sein, enthält einen Verstoß gegen die ärztlichen Standespflichten, *Ministerial-Blatt für Medizinal- und medizinische Unterrichtsangelegenheiten* 3, p. 217.

Preußischer Ehrengerichtshof für Ärzte (1907) Beschluß vom 8. Januar 1907, *Ministerial-Blatt für Medizinal- und medizinische Unterrichtsangelegenheiten* 7, pp. 239–240.

Preußischer Ehrengerichtshof für Ärzte (1910) Urteil vom 5. April 1910, *Ministerial-Blatt für Medizinal- und medizinische Unterrichtsangelegenheiten* 10, p. 376.

Preußischer Ehrengerichtshof für Ärzte (1908–34) *Entscheidungen des Preußischen Ehrengerichtshofes für Ärzte. Im Auftrage des Ehrengerichtshofes herausgegeben*, 5 vols, Berlin, Richard Schoetz.

Rapmund, O. and Dietrich, E. (eds) (1898/99) *Ärztliche Rechts- und Gesetzkunde*, Leipzig, Georg Thieme.

Reichsgericht (1880) Ist die an einem Einwilligenden begangene Körperverletzung strafbar? (I. Strafsenat, Urteil v. 15. November 1880), in *Entscheidungen des Reichsgerichts. Herausgegeben von den Mitgliedern des Gerichtshofes. Entscheidungen in Strafsachen*, Leipzig, Veit & Comp., vol. 2, pp. 442–443.

Reichsgericht (1882) Wird die Strafbarkeit von Körperverletzungen oder Tötungen dadurch ausgeschlossen, daß festgestellt wird, dieselben seien in einem nach vereinbarten oder herkömmlichen Regeln stattgehabten, nach Beschaffenheit der dabei gebrauchten, nicht tödlichen Waffen aber nicht strafbaren Zweikampfe verübt worden? (III. Strafsenat, Urteil v. 22. Februar 1882), in *Entscheidungen des Reichsgerichts. Herausgegeben von den Mitgliedern des Gerichtshofes. Entscheidungen in Strafsachen*, Leipzig, Veit & Comp., vol. 6, pp. 61–64.

Reichsgericht (1886) Welchen Sinn haben im § 300 St.G.B.'s die Thatbestandsmerkmale 'Privatgeheimnis' und 'anvertrauen'? Wem steht bei dem Vergehen gegen § 300 a. a. O. das Recht zu, den Strafantrag zu stellen, insbesondere hinsichtlich der Verletzung ärztlicher Geheimnisse der Ehefrau? (III. Strafsenat, Urteil v. 22. Oktober 1885), in *Entscheidungen des Reichsgerichts. Herausgegeben von den Mitgliedern des Gerichtshofes. Entscheidungen in Strafsachen*, Leipzig, Veit & Comp., vol. 13, pp. 60–64.

Reichsgericht (1889) Ist die Vernehmung eines behandelnden Arztes von vorgängiger Entbindung von der Verpflichtung zur Amtsverschwiegenheit seitens des Patienten abhängig? St. P. O. § 52 Nr. 3 (I. Strafsenat, Urteil v. 8. Juli 1889), in *Entscheidungen des Reichsgerichts. Herausgegeben von den Mitgliedern des Gerichtshofes. Entscheidungen in Strafsachen*, Leipzig, Veit & Comp., vol. 19, pp. 364–367.

Reichsgericht (1894a) Sind die Wahrnehmungen, welche ein Arzt bei der Untersuchung einer wegen erlittener Mißhandlung ihn konsultierenden Person macht, Privatgeheimnisse, die ihm anvertraut sind, selbst wenn der Patient über die Mißhandlung schon anderen Personen Mitteilung gemacht hat? Ist es als ein 'Offenbaren' anzusehen, wenn der Arzt unbestimmte Gerüchte, welche über die Mißhandlung verbreitet sind, als richtig bestätigt? (IV. Strafsenat, Urteil v. 26. Juni 1894), in *Entscheidungen des Reichsgerichts. Herausgegeben von den Mitgliedern des Gerichtshofes und der Reichsanwaltschaft. Entscheidungen in Strafsachen*, Leipzig, Veit & Comp., vol. 26, pp. 5–8.

Reichsgericht (1894b) Von welchen rechtlichen Voraussetzungen hängt die Strafbarkeit oder Straflosigkeit von Körperverletzungen ab, welche zum Zwecke des Heilverfahrens von Ärzten bei operativen Eingriffen begangen werden? (III. Strafsenat, Urteil v. 31. Mai 1894), in *Entscheidungen des Reichsgerichts. Herausgegeben von den Mitgliedern des Gerichtshofes und der Reichsanwaltschaft. Entscheidungen in Strafsachen*, Leipzig, Veit & Comp., vol. 25, pp. 375–389.

Reichsgericht (1903) Inwieweit muß das Recht eines Arztes, nach Maßgabe von § 383 Abs. 1 Ziff. 5 C.P.O. sein Zeugnis zu verweigern, hinter einer höheren sittlichen Pflicht zurücktreten? (VI. Civilsenat, Beschluß v. 19. Januar 1903), in *Entscheidungen des Reichsgerichts. Herausgegeben von den Mitgliedern des Gerichtshofes und der Reichsanwaltschaft. Entscheidungen in Civilsachen*, Leipzig, Veit & Comp., vol. 53, pp. 315–319.

Reichsgericht (1905a) 1. Kann eine Befugnis des Arztes zur Offenbarung von Privatgeheimnissen, die ihm kraft seines Standes oder Gewerbes anvertraut sind, auch durch anderweite Berufspflichten des Arztes begründet werden? 2. Unter welchen Voraussetzungen ist anzunehmen, daß eine anvertraute Tatsache ihre Eigenschaft als Privatgeheimnis durch anderweite Kundgebung verloren habe? (II. Strafsenat, Urteil v. 16. Mai 1905), in *Entscheidungen des Reichsgerichts. Herausgegeben von den Mitgliedern des Gerichtshofes und der Reichsanwaltschaft. Entscheidungen des Reichsgerichts in Strafsachen*, Leipzig, Veit & Comp., vol. 38, pp. 62–66.

Reichsgericht (1905b) Strafbarkeit von Körperverletzungen, welche zum Zwecke des Heilverfahrens von nicht wissenschaftlich gebildeten Heilkundigen bei operativen Eingriffen begangen werden (III. Strafsenat, Urteil v. 10. April 1905), in *Entscheidungen des Reichsgerichts. Herausgegeben von den Mitgliedern des Gerichtshofes und der Reichsanwaltschaft. Entscheidungen in Strafsachen*, Leipzig, Veit & Comp., vol. 38, pp. 34–37.

Reichsgericht (1908) Nach welchen Grundsätzen bestimmt sich die Schadensersatzpflicht des Arztes, der an einem Kinde ohne Einwilligung des gesetzlichen Vertreters eine Operation vorgenommen hat? Darf der Leiter einer größeren Klinik seinem Personal die Verhandlungen wegen Zustimmung der Einwilligungsberechtigten überlassen? Inwiefern kann die Einwilligung vermutet werden? (VI. Zivilsenat, Urteil v. 27. Mai 1908), in *Entscheidungen des Reichsgerichts. Herausgegeben von den Mitgliedern des Gerichtshofes und der Reichsanwaltschaft. Entscheidungen in Zivilsachen. Neue Folge*, Leipzig, Veit & Comp., vol. 18, pp. 431–438.

Reichsgericht (1912) 1. Ist der Arzt verpflichtet, den Kranken auf die nachteiligen Folgen aufmerksam zu machen, die möglicherweise bei einer beabsichtigten Operation entstehen können? 2. Zur Frage der Beweislast beim Eintritte schädlicher Folgen einer Operation (III. Zivilsenat, Urteil v. 1. März 1912), in *Entscheidungen des Reichsgerichts. Herausgegeben von den Mitgliedern des Gerichtshofes und der Reichsanwaltschaft. Entscheidungen in Zivilsachen. Neue Folge,* Leipzig, Veit & Comp., vol. 28, pp. 432–436.

Reichsgericht (1914) 1. Hat der Angeklagte einen Revisionsgrund, wenn der als Zeuge vernommene Arzt über sein Zeugnisverweigerungsrecht aus § 52 StPO. vom Vorsitzenden unrichtig belehrt worden war? 2. Ist der nach § 52 zur Zeugnisverweigerung berechtigte Zeuge verpflichtet, das ihm anvertraute Geheimnis preiszugeben, wenn er erklärt hat, von seinem Verweigerungsrecht keinen Gebrauch zu machen, oder wenn er seine Aussage bereits gemacht hat, die jedenfalls zum Teil von seinem Verweigerungsrecht betroffen wurde? 3. Unterschied von unvollständiger und zufolge Verschweigens unrichtiger Zeugenaussage (V. Strafsenat, Urteil v. 16. Mai 1914), in *Entscheidungen des Reichsgerichts. Herausgegeben von den Mitgliedern des Gerichtshofes und der Reichsanwaltschaft. Entscheidungen in Strafsachen,* Leipzig, Veit & Comp., vol. 48, pp. 269–274.

Reichsgericht (1936) 1. Ist ein Arzt zu körperlichen Eingriffen auch ohne oder sogar gegen den Willen des Kranken befugt? 2. Handelt ein Arzt bei einem gegen den Willen des Kranken vorgenommenen Eingriff nicht schuldhaft, wenn er glaubt, dazu auch ohne Einwilligung des Betroffenen rechtlich befugt zu sein? 3. Kann das Gericht schriftliche Äußerungen von Ärzten berücksichtigen, die der gerichtliche Sachverständige herbeigezogen und seinem Gutachten beigefügt hat? (III. Zivilsenat, Urteil v. 19. Juni 1936), in *Entscheidungen des Reichsgerichts. Herausgegeben von den Mitgliedern des Gerichtshofes und der Reichsanwaltschaft. Entscheidungen in Zivilsachen,* Berlin and Leipzig, Walter de Gruyter & Co., vol. 151, pp. 349–357.

Reichsgericht (1940) 1. Zur Sorgfaltpflicht des Arztes vor einem schwerwiegenden Eingriff. 2. Wie ist die Rechtslage, wenn der Arzt bewußtermaßen einen schwerwiegenden Eingriff ohne die Einwilligung des Kranken vorgenommen hat, obwohl ihm deren Einholung möglich gewesen wäre, und wenn sich der Eingriff später als nicht erforderlich herausstellte? (III. Zivilsenat, Urteil v. 8. März 1940), in *Entscheidungen des Reichsgerichts. Herausgegeben von den Mitgliedern des Gerichtshofes und der Reichsanwaltschaft. Entscheidungen in Zivilsachen,* Berlin, Walter de Gruyter & Co., vol. 163, pp. 129–139.

Rosenberg, W. (1903) Strafbare Heilungen, *Der Gerichtssaal* 62, pp. 62–83.

Rudeck, W. (1902) *Medizin und Recht. Geschlechtsleben und –Krankheiten in medizinisch-juristisch-kulturgeschichtlicher Bedeutung,* Berlin, H. Barsdorf.

Runge, F. (1881) Aerztliche Ehrengerichte, *Ärztliches Vereinsblatt für Deutschland* 8, cols 7–8.

Rutkowsky (Anwaltsassessor, Berlin) (1938) Die ärztliche Schweigepflicht als Hindernis der Wahrheitsfindung, *Deutsches Recht* 8, pp. 430–431.

Sauter, F. (1910) *Das Berufsgeheimnis und sein strafrechtlicher Schutz. (§ 300 R.St.G.B.),* Breslau, Schletter'sche Buchhandlung.

Schäfer, K. (1936) Zu § 13 der Ärzteordnung (Berufsverschwiegenheit), *Deutsche Justiz* 98, pp. 374–376.

Schäfer, K. (1937) Bemerkungen zur ärztlichen Schweigepflicht, *Deutsches Strafrecht,* Neue Folge 4, pp. 197–200.

Schlegtendal (Kreisphysikus in Lennep) (1895) Das Berufsgeheimnis der Aerzte, *Deutsche Medicinische Wochenschrift* 21, pp. 503–506.

Schmidt, H. [no date; *c.* 1884] *Streif-Lichter über die Stellung des Arztes in der Gegenwart und sein Verhältnis zur Praxis oder Die Medizin was sie ist, was sie kann und was sie will*, Berlin, Otto Dreyer.

Schmidt, H. (1907) *Das ärztliche Berufsgeheimnis*, Jur. Diss. Leipzig, Jena, Gustav Fischer.

Schmidt, R. (1900) *Die strafrechtliche Verantwortlichkeit des Arztes für verletzende Eingriffe. Ein Beitrag zur Lehre der Straf- und Schuldausschliessungsgründe*, Jena, Gustav Fischer.

Scholz, F. (1914) *Von Ärzten und Patienten: Lustige und unlustige Plaudereien. Mit dem Bildnis des Verfassers und Originalfederzeichnungen von O. Merté*, 4th edn, Munich, Verlag der Ärztlichen Rundschau Otto Gmelin.

Scholz, F. (1927) *Von Aerzten und Patienten. Lustige und unlustige Plaudereien*, 5th edn, in E. Liek (ed.), Munich, Verlag der Aerztlichen Rundschau Otto Gmelin.

Schumacher, W. (1931) *Das ärztliche Berufsgeheimnis nach § 300 RStGB*, Berlin, Richard Schoetz.

Schweninger, E. (1906) *Der Arzt*, Frankfurt/Main, Literarische Anstalt Rütten & Loening.

Seelig and Hacke (Rechtsanwälte beim Reichsgericht) (1911) Vom Reichsgericht. Bürgerliches Gesetzbuch. 2. § 107 BGB. Operation eines Minderjährigen ohne Einwilligung des gesetzlichen Vertreters (Urteil v. 30. Juni 1911), *Juristische Wochenschrift* 40, pp. 747–749.

Seelig and Scheele (Rechtsanwälte beim Reichsgericht) (1907) Vom Reichsgericht. 1. Reichsrecht. Bürgerliches Gesetzbuch. 2. §§ 106, 107, 113, 276, 287 BGB. Begründet Vornahme der Operation ohne Zustimmung des Patienten unter allen Umständen ein Verschulden des Arztes? (Urteil v. 21. Juni 1907), *Juristische Wochenschrift* 36, pp. 505–506.

Seréxhe, L. (1906) *Die Verletzung fremder Geheimnisse*, Jur. Diss. Freiburg i. Brsg., Karlsruhe, G. Braunsche Hofbuchdruckerei.

Simmel, Georg (1958) *Soziologie. Untersuchungen über die Formen der Vergesellschaftung* [1st edn 1908], 4th edn, Berlin, Duncker & Humblot.

Simonson (Oberlandesgerichtsrat, Breslau) (1904) Das Berufsgeheimnis der Aerzte und deren Recht der Zeugnisverweigerung, *Deutsche Juristen-Zeitung* 9, cols 1014–1017.

Spinner, J. R. (1914) *Ärztliches Recht. Unter besonderer Berücksichtigung deutschen, schweizerischen, österreichischen und französischen Rechts*, Berlin, Julius Springer.

Stenglein, M. (1899a) Operationsrecht des Chirurgen, *Deutsche Juristen-Zeitung* 4, p. 151.

Stenglein, M. (1899b) Das Recht des Arztes, zu operiren, *Münchener Medicinische Wochenschrift* 46, pp. 525–527.

Stenglein, M. (1902) Literarische Anzeigen: Albert Moll, Aerztliche Ethik, *Der Gerichtssaal* 60, pp. 383–384.

[Stooss, C.] (1892) Literatur-Anzeigen.–Bibliographie: Dr. L. Oppenheim, Professor des Strafrechts an der Universität Basel. Das ärztliche Recht zu körperlichen Eingriffen an Kranken und Gesunden. B. Schwabe, Basel 1892, *Zeitschrift für Schweizer Strafrecht* 5, p. 465.

Stooss, C. (1893) Operativer Eingriff und Körperverletzung, *Zeitschrift für Schweizer Strafrecht* 6, pp. 53–61.

Stooss, C. (1894) Ist der ärztliche Zweck das gewohnheitsrechtliche Fundament zum ärztlichen Eingriff? *Zeitschrift für Schweizer Strafrecht* 7, pp. 192–198.

Stooss, C. (1898) *Chirurgische Operation und ärztliche Behandlung. Eine strafrechtliche Studie*, Berlin, Otto Liebmann.

Stooss, C. (1899a) Die strafrechtliche Natur ärztlicher Handlungen, *Deutsche Juristen-Zeitung* 4, pp. 184–186.

Stooss, C. (1899b) Die ärztliche Behandlung im Strafrecht, *Schweizerische Zeitschrift für Strafrecht* 12, pp. 24–32.

Stooss, C. (1899c) Die strafrechtliche Natur der ärztlichen Behandlung, *Deutsche Medicinische Wochenschrift* 25, pp. 247–248.

Stooss, C. (1902) Aerztliche Behandlung und Körperverletzung, *Deutsche Juristen-Zeitung* 7, pp. 566–568.

Stooss, C. (1903) Aerztliche Behandlung und Körperverletzung, *Berliner Aerzte-Correspondenz* 8, pp. 6–7.

Styrap, J. de (1878) *A Code of Medical Ethics*, London, Churchill. Reprinted in R. Baker (ed.) (1995), *The Codification of Medical Morality*, vol. 2: *Anglo-American Medical Ethics and Medical Jurisprudence in the Nineteenth Century*, Dordrecht, Kluwer Academic Publishers, pp. 149–171.

Thiersch, J. (1894) Sind chirurgische Operationen und operative Eingriffe körperliche Misshandlungen im strafrechtlichen Sinne?, *Ärztliches Vereinsblatt für Deutschland* 23, cols 473–479.

Thilo, G. (1864) *Die Preußische Disziplinargesetzgebung für die unmittelbaren und mittelbaren Staatsbeamten*, Berlin, J. Guttentag.

Tränkner (Oberlandesgerichtsrat, Dresden) (1899) Oberlandesgericht Dresden. Darf ein Arzt ohne Genehmigung des Patienten eine Operation vornehmen?, *Deutsche Juristen-Zeitung* 4, pp. 199–200.

Ughetti, J. B. (1899) *Zwischen Ärzten und Clienten. Erinnerungen eines alten Arztes*, transl. by G. Galli, Vienna and Leipzig, Wilhelm Braumüller.

Vierordt, H. (1893) Arzt und Patient, *Deutsche Revue über das gesamte nationale Leben der Gegenwart* 18/3, pp. 97–114.

Vogler, E. (1939) *Betrachtungen zur Begrenzung der ärztlichen Schweigepflicht nach § 13 Abs. 3 der Reichsärzteordnung*, Jur. Diss. Marburg, Düsseldorf, G. H. Nolte.

Vollmann (1916) Umschau. Beratungsstellen – Berufsgeheimnis – Kurpfuscherei – ärztliche Ausbildung, *Ärztliches Vereinsblatt für Deutschland* 43, cols 455–462.

[Wallichs, J.] (1894) Medicin und Rechtspflege, *Ärztliches Vereinsblatt für Deutschland* 23, cols 497–499.

Wallichs, J. (1895a) Eduard Graf †, *Ärztliches Vereinsblatt für Deutschland* 22 (1895), cols 473–475.

Wallichs, J. (1895b) Dr. Eduard Graf's äusserer Lebensgang, *Ärztliches Vereinsblatt für Deutschland* 22 (1895), cols 489–491.

Weizmann, H. (1909) *Das Berufsgeheimnis. (§ 300 RStGB.)*, Jur. Diss. Breslau, Borna-Leipzig, Buchdruckerei Robert Noske.

Wolff, J. (1896) *Der Praktische Arzt und sein Beruf. Vademecum für angehende Praktiker*, Stuttgart, Ferdinand Enke.

Ziemssen, H. von (1887) Der Arzt und die Aufgaben des ärztlichen Berufes, in H. von Ziemssen (ed.), *Klinische Vorträge*, Leipzig, F. C. W. Vogel, pp. 1–23.

Ziemssen, O. (1898) Ueber den medicinisch-klinischen Unterricht. Discussion, in *Verhandlungen des Congresses für Innere Medicin*, vol. 16, pp. 68–69.

Ziemssen, O. (1899) *Die Ethik des Arztes als Medicinischer Lehrgegenstand*, Leipzig, Georg Thieme.

Zitelmann, E. (1907) Die Haftung des Arztes aus ärztlicher Behandlung, *Deutsche Medizinische Wochenschrift* 33, pp. 2052–2054, 2098–2101, 2144–2147.

Zschok, W. (1903) *§ 300 StrGB*, Jur. Diss. Rostock, Carl Hinstorffs Buchdruckerei.

Secondary literature

Annas, G. J. and Grodin, M. A. (eds) (1992) *The Nazi Doctors and the Nuremberg Code. Human Rights in Human Experimentation*, New York and Oxford, Oxford University Press.

Baker, R. (1995a) The Historical Context of the American Medical Association's 1847 *Code of Ethics*, in R. Baker (ed.), *The Codification of Medical Morality*, vol. 2: *Anglo-American Medical Ethics and Medical Jurisprudence in the Nineteenth Century*, Dordrecht, Kluwer Academic Publishers, pp. 47–63.

Baker, R. (ed.) (1995b) *The Codification of Medical Morality*, vol. 2: *Anglo-American Medical Ethics and Medical Jurisprudence in the Nineteenth Century*, Dordrecht, Kluwer Academic Publishers.

Baker, R. B., Caplan, A. L., Emanuel, L. L. and Latham, S. L. (eds) (1999) *The American Medical Ethics Revolution: How the AMA's Code of Ethics Has Transformed Physicians' Relationships to Patients, Professionals, and Society*, Baltimore, MD and London, Johns Hopkins University Press.

Baker, R., Porter, D. and Porter, R. (eds) (1993) *The Codification of Medical Morality*, vol. 1: *Medical Ethics and Etiquette in the Eighteenth Century*, Dordrecht, Kluwer Academic Publishers.

Bartrip, P. (1995) An Introduction to Jukes Styrap's *A Code of Medical Ethics* (1878), in R. Baker (ed.), *The Codification of Medical Morality*, vol. 2: *Anglo-American Medical Ethics and Medical Jurisprudence in the Nineteenth Century*, Dordrecht, Kluwer Academic Publishers, pp. 145–171.

Bartrip, P. (1996) *Themselves Writ Large: The British Medical Association 1832–1966*, London, BMJ.

Beauchamp, T. L. (1995) Worthington Hooker on Ethics in Clinical Medicine, in R. Baker (ed.), *The Codification of Medical Morality*, vol. 2: *Anglo-American Medical Ethics and Medical Jurisprudence in the Nineteenth Century*, Dordrecht, Kluwer Academic Publishers, pp. 105–119.

Benzenhöfer, U. (1999) *Der gute Tod? Euthanasie und Sterbehilfe in Geschichte und Gegenwart*, Munich, C. H. Beck.

Benzenhöfer, U. (2006) *Zur Genese des Gesetzes zur Verhütung erbkranken Nachwuchses*, Münster, Klemm & Oelschläger.

Berg, M. and Cocks, G. (eds) (1997) *Medicine and Modernity: Public Health and Medical Care in Nineteenth- and Twentieth-Century Germany*, Cambridge, Cambridge University Press.

Berlant, J. L. (1975) *Profession and Monopoly: A Study of Medicine in the United States and Great Britain*, Berkeley, CA, University of California Press.

Binder, J. (2000) *Zwischen Standesrecht und Marktwirtschaft. Ärztliche Werbung zu Beginn des 20. Jahrhunderts im deutsch-englischen Vergleich*, Frankfurt/ Main, Peter Lang.

Bleker, J. (1996) Das Ende des männlichen Berufsmonopols in Deutschland. Die ersten 'legitimen weiblichen Ärzte' werden approbiert, in H. Schott (ed.), *Meilensteine der Medizin*, Dortmund, Harenberg Verlag, pp. 396–402.

Bleker, J. and Jachertz, N. (1993) *Medizin im 'Dritten Reich'*, 2nd edn, Cologne, Deutscher Ärzte-Verlag.

Bock, G. (1997) Sterilization and 'Medical' Massacres in National Socialist Germany: Ethics, Politics, and the Law, in M. Berg and G. Cocks (eds), *Medicine and Modernity: Public Health and Medical Care in Nineteenth- and Twentieth-Century Germany*, Cambridge, Cambridge University Press, pp. 149–172.

Bockelmann, P. (1981) Der ärztliche Heileingriff in Beiträgen zur Zeitschrift für die gesamte Strafrechtswissenschaft im ersten Jahrhundert ihres Bestehens, *Zeitschrift für die gesamte Strafrechtswissenschaft* 93, pp. 105–150.

Bourdieu, P. (1972) *Esquisse d'une théorie de la pratique: precede de trois études d'ethnologie kabyle*, Geneva, Droz.

Brand, U. (1977) *Ärztliche Ethik im 19. Jahrhundert. Der Wandel ethischer Inhalte im medizinischen Schrifttum*, Freiburg/Breisgau, Hans Ferdinand Schulz.

Burleigh, M. (1994) *Death and Deliverance: 'Euthanasia' in Germany c. 1900– 1945*, Cambridge, Cambridge University Press.

Burnham, J. C. (1998) *How the Idea of Profession Changed the Writing of Medical History* (*Medical History*, Supplement No. 18), London, Wellcome Institute for the History of Medicine.

Cane, W. (1952) 'Medical Euthanasia'. A Paper, Published in Latin in 1826, Translated and Reintroduced to the Medical Profession, *Journal of the History of Medicine and Allied Sciences* 7, pp. 401–416.

Cario, D. (1999) *Albert Moll (1862–1939). Leben, Werk und Bedeutung für die Medizinische Psychologie*, MD thesis, Mainz.

Dinges, M. (ed.) (1996) *Medizinkritische Bewegungen im Deutschen Reich (ca. 1870 – ca. 1933)*, Stuttgart, Franz Steiner Verlag.

Drees, A. (1988) *Die Ärzte auf dem Weg zu Prestige und Wohlstand: Sozialgeschichte der württembergischen Ärzte im 19. Jahrhundert*, Münster, Coppenrath.

Eben, A. K. (1998) *Medizinische Ethik im weltanschaulich-religiösen Kontext: Albert Moll und Albert Niedermeyer im Vergleich*, MD thesis, Munich, Verlag Uni-Druck.

Eckart, W. U. (1997) *Medizin und Kolonialimperialismus: Deutschland 1884– 1945*, Paderborn, Munich, Vienna and Zurich, Ferdinand Schöningh.

Eckart, W. U. and Gradmann, C. (eds) (1996) *Die Medizin und der Erste Weltkrieg*, Paffenweiler, Centaurus-Verlagsgesellschaft.

Eckart, W. U. and Reuland, A. (2006) First Principles: Julius Moses and Medical Experimentation in the Late Weimar Republic, in W. U. Eckart (ed.), *Man, Medicine and the State. The Human Body as an Object of Government Sponsored Medical Research in the 20th Century*, Stuttgart, Franz Steiner Verlag, pp. 35–47.

Edelstein, L. (1967) *Ancient Medicine: Selected Papers of Ludwig Edelstein*, in O. Temkin and C. L. Temkin (eds), Baltimore, MD, Johns Hopkins Press.

Elkeles, B. (1985) Medizinische Menschenversuche gegen Ende des 19. Jahrhunderts und der Fall Neisser, *Medizinhistorisches Journal* 20, pp. 135–148.

Elkeles, B. (1989) Die schweigsame Welt von Arzt und Patient. Einwilligung und Aufklärung in der Arzt-Patienten-Beziehung des 19. und frühen 20. Jahrhunderts, *Medizin, Gesellschaft und Geschichte* 8, pp. 63–91.

Elkeles, B. (1996a) *Der moralische Diskurs über das medizinische Menschenexperiment im 19. Jahrhundert*, Stuttgart, Jena and New York, Gustav Fischer.

Elkeles, B. (1996b) Der Patient und das Krankenhaus, in A. Labisch and R. Spree (eds), *'Einem jeden Kranken in einem Hospitale sein eigenes Bett'. Zur Sozialgeschichte des Allgemeinen Krankenhauses in Deutschland im 19. Jahrhundert*, Frankfurt/Main, Campus, pp. 357–373.

Elkeles, B. (2001) Wissenschaft, Medizinethik und gesellschaftliches Umfeld: Die Diskussion um den Heilversuch um 1900, in A. Frewer and J. N. Neumann (eds), *Medizingeschichte und Medizinethik: Kontroversen und Begründungsansätze 1900–1950*, Frankfurt and New York, Campus Verlag, pp. 21–43.

Elkeles, B. (2004) The German Debate on Human Experimentation between 1880 and 1914, in V. Roelcke and G. Maio (eds), *Twentieth Century Ethics of Human Subjects Research. Historical Perspectives on Values, Practices and Regulations*, Stuttgart, Franz Steiner Verlag, 2004, pp. 19–33.

Engel, J. (1965) *Ottomar Rosenbach*, Zurich, Juris-Verlag.

Engelhardt, D. von (1989) Entwicklung der ärztlichen Ethik im 19. Jahrhundert – medizinische Motivation und gesellschaftliche Legitimation, in A. Labisch and R. Spree (eds), *Medizinische Deutungsmacht im sozialen Wandel des 19. und 20. Jahrhunderts*, Bonn, Psychiatrie-Verlag, pp. 75–88.

Engelhardt, D. von (1996) Wahrheit am Krankenbett im geschichtlichen Überblick, *Schweizerische Rundschau für Medizin (Praxis)* 85, pp. 432–439.

Engelhardt, D. von (2000) *Euthanasie zwischen Lebensverkürzung und Sterbebeistand. Vortrag gehalten vor der Juristischen Gesellschaft Mittelfranken zu Nürnberg e. V. am 22. November 1999*, Regensburg, S. Roderer Verlag.

Engstrom, E. J. (2003) *Clinical Psychiatry in Imperial Germany: A History of Psychiatric Practice*, Ithaca, NY and London, Cornell University Press.

Evans, R. J. (1987) *Death in Hamburg: Society and Politics in the Cholera Years 1830–1910*, Oxford, Clarendon.

Featherstone, M. (1991) Georg Simmel: An Introduction, *Theory, Culture & Society* 8, pp. 1–16.

Ferguson, A. H. (2005) *Should a Doctor Tell? Medical Confidentiality in Interwar England and Scotland*, PhD thesis, University of Glasgow.

French, R. (1993) The Medical Ethics of Gabriele de Zerbi, in A. Wear, J. Geyer-Kordesch and R. French (eds), *Doctors and Ethics: The Earlier Historical Setting of Professional Ethics*, Amsterdam and Atlanta, GA: Rodopi, pp. 72–97.

Frevert, U. (1984) *Krankheit als politisches Problem 1770–1880. Soziale Unterschichten in Preußen zwischen medizinischer Polizei und staatlicher Sozialversicherung*, Göttingen, Vandenhoeck & Ruprecht.

Frevert, U. (1995) *Men of Honour: A Social and Cultural History of the Duel*, translated by A. Williams, Cambridge, Polity Press.

Frewer, A. and Roelcke, V. (eds) (2001) *Die Institutionalisierung der Medizinhistoriographie. Entwicklungslinien vom 19. ins 20. Jahrhundert*, Stuttgart, Franz Steiner Verlag.

Frisby, D. (2002) *Georg Simmel. Revised Edition*, London and New York, Routledge.

Göckenjan, G. (1985) *Kurieren und Staat machen. Gesundheit und Medizin in der bürgerlichen Welt*, Frankfurt/Main, Suhrkamp.

Göckenjan, G. (1989) Wandlungen im Selbstbild des Arztes seit dem 19. Jahrhundert, in A. Labisch and R. Spree (eds), *Medizinische Deutungsmacht im sozialen Wandel des 19. und 20. Jahrhunderts*, Bonn, Psychiatrie-Verlag, pp. 89–102.

Gradmann, C. (2001) Robert Koch and the Pressures of Scientific Research: Tuberculosis and Tuberculin, *Medical History* 45, pp. 1–32.

Gradmann, C. (2005) *Krankheit im Labor: Robert Koch und die medizinische Bakteriologie*, Göttingen, Wallstein.

Grossmann, A. (1995) *Reforming Sex. The German Movement for Birth Control and Abortion Reform, 1920–1950*, New York and Oxford, Oxford University Press.

Haakonssen, L. (1997) *Medicine and Morals in the Enlightenment: John Gregory, Thomas Percival and Benjamin Rush*, Amsterdam, Rodopi.

Hahn, S. (1984) Die ärztliche Ethik im Leben eines Arztes der Seele – Überlegungen zur medizinisch-ethischen Konzeption Albert Molls (1862–1939), *Zeitschrift für die gesamte innere Medizin* 39, pp. 558–561.

Hau, M. (2001) Experten für Menschlichkeit? Ärztliche Berufsethik, Lebensreform und die Krise der Medizin in der Weimarer Republik, in A. Frewer and J. N. Neumann (eds), *Medizingeschichte und Medizinethik: Kontroversen und Begründungsansätze 1900–1950*, Frankfurt and New York, Campus Verlag, pp. 124–142.

Hermann, C. (1929) *Max Dessoir. Mensch und Werk*, Stuttgart, Ferdinand Enke.

Herold-Schmidt, H. (1997) Ärztliche Interessenvertretung im Kaiserreich 1871–1914, in R. Jütte (ed.), *Geschichte der deutschen Ärzteschaft. Organisierte Berufs– und Gesundheitspolitik im 19. und 20. Jahrhundert*, Cologne, Deutscher Ärzte-Verlag, pp. 43–95.

Hochtritt, H. G. (1969) *Die Berufsgerichtsbarkeit. Eine verfassungsrechtliche Untersuchung über die Verfassung und das Verfahren der Berufsgerichte*, Cologne and Berlin, Deutscher Ärzte-Verlag.

Holl, K. (2007) *Ludwig Quidde (1858–1941). Eine Biografie*, Düsseldorf, Droste Verlag.

Huerkamp, C. (1985) *Der Aufstieg der Ärzte im 19. Jahrhundert. Vom gelehrten Stand zum professionellen Experten: Das Beispiel Preußens*, Göttingen, Vandenhoeck & Ruprecht.

Huerkamp, C. (1990) The Making of the Modern Medical Profession, 1800–1914: Prussian Doctors in the Nineteenth Century, in G. Cocks and K. H. Jarausch (eds), *German Professions, 1800–1950*, New York and Oxford, Oxford University Press, pp. 66–84.

Jütte, R. (ed.) (1997a) *Geschichte der deutschen Ärzteschaft. Organisierte Berufs– und Gesundheitspolitik im 19. und 20. Jahrhundert*, Cologne, Deutscher Ärzte-Verlag.

Jütte, R. (1997b) Die Entwicklung des ärztlichen Vereinswesens und des organisierten Ärztestandes bis 1871, in R. Jütte (ed.), *Geschichte der deutschen Ärzteschaft. Organisierte Berufs– und Gesundheitspolitik im 19. und 20. Jahrhundert*, Cologne, Deutscher Ärzte-Verlag, pp. 15–42.

Karstens, K. (1984) *'Déontologie médicale' im 19. Jahrhundert*, MD thesis, Albert-Ludwigs-Universität Freiburg im Breisgau.

Kater, M. H. (1985) Professionalization and Socialization of Physicians in Wilhelmine and Weimar Germany, *Journal of Contemporary History* 20, pp. 677–701.

Kater, M. H. (1989) *Doctors under Hitler*, Chapel Hill, NC and London, The University of North Carolina Press.

Kater, M. H. (1990) Die Medizin im nationalsozialistischen Deutschland und Erwin Liek, *Geschichte und Gesellschaft* 16, pp. 440–463.

Katz, J. (1986) *The Silent World of Doctor and Patient*, New York, The Free Press.

Knüpling, H. (1965) *Untersuchungen zur Vorgeschichte der Deutschen Ärzteordnung von 1935*, MD thesis, Freie Universität Berlin.

Koch, C. (2004) *Schwangerschaftsabbruch (§§ 218ff. StGB). Reformdiskussion und Gesetzgebung von 1870 bis 1945*, Münster, Lit Verlag.

Kocka, J. (1988) German History before Hitler: The Debate about the German Sonderweg, *Journal of Contemporary History* 23, pp. 3–16.

Kurzweg, A. (1976) *Die Geschichte der Berliner 'Gesellschaft für Experimental-Psychologie' mit besonderer Berücksichtigung ihrer Ausgangssituation und des Wirkens von Max Dessoir*, MD thesis, Freie Universität Berlin.

Labisch, A. (1997) From Traditional Individualism to Collective Professionalism: State, Patient, Compulsory Health Insurance, and the Panel Doctor Question in Germany, 1883–1931', in M. Berg and G. Cocks (eds), *Medicine and Modernity: Public Health and Medical Care in Nineteenth- and Twentieth-Century Germany*, Cambridge, Cambridge University Press, pp. 35–54.

Leven, K.-H. and Prüll, C.-R. (eds) (1994) *Selbstbilder des Arztes im 20. Jahrhundert. Medizinhistorische und medizinethische Aspekte*, Freiburg im Brsg., Hans Ferdinand Schulz Verlag.

Lifton, R. J. (1986) *The Nazi Doctors: Medical Killing and the Psychology of Genocide*, New York, Basic Books.

Linden, D. E. J. (1999) Gabriele de Zerbi's *De cautelis medicorum* and the Tradition of Medical Prudence, *Bulletin of the History of Medicine* 73, pp. 19–37.

Luther, E. (1975) Die Herausbildung und gesellschaftliche Sanktionierung der ärztlichen Standesauffassung in der zweiten Hälfte des 19. Jahrhunderts,

Wissenschaftliche Zeitschrift der Universität Halle, Math.-Nat. Reihe 24, H. 2, pp. 5–28.

Maehle, A.-H. (1990) Der Streit über das Reichsimpfgesetz von 1874, *Medizin, Gesellschaft und Geschichte* 9, pp. 127–148.

Maehle, A.-H. (1996) Organisierte Tierversuchsgegner: Gründe und Grenzen ihrer gesellschaftlichen Wirkung, 1879–1933, in M. Dinges (ed.), *Medizinkritische Bewegungen im Deutschen Reich (ca. 1870 – ca. 1933),* Stuttgart, Franz Steiner Verlag, pp. 109–125.

Maehle, A.-H. (1999) Professional Ethics and Discipline: The Prussian Medical Courts of Honour, 1899–1920, *Medizinhistorisches Journal* 34, pp. 309–338.

Maehle, A.-H. (2000) Assault and Battery, or Legitimate Treatment? German Legal Debates on the Status of Medical Interventions without Consent, c. 1890–1914, *Gesnerus* 57, pp. 206–221.

Maehle, A.-H. (2001) Zwischen medizinischem Paternalismus und Patientenautonomie: Albert Molls 'Ärztliche Ethik' (1902) im historischen Kontext, in A. Frewer and J. N. Neumann (eds), *Medizingeschichte und Medizinethik: Kontroversen und Begründungsansätze 1900–1950,* Frankfurt and New York, Campus Verlag, pp. 44–56.

Maehle, A.-H. (2003a) Protecting Patient Privacy or Serving Public Interests? Challenges to Medical Confidentiality in Imperial Germany, *Social History of Medicine* 16, pp. 383–401.

Maehle, A.-H. (2003b) Ärztlicher Eingriff und Körperverletzung: Zu den historisch-rechtlichen Wurzeln des Informed Consent in der Chirurgie, 1892–1940, *Würzburger medizinhistorische Mitteilungen* 22, pp. 178–187.

Maehle, A.-H. (2007) Albert Moll (1862–1939), in W. F. Bynum and H. Bynum (eds), *Dictionary of Medical Biography,* Westport, CT and London, Greenwood Press, vol. 4, pp. 884–885.

Martin, B. and Szelény, I. (2000) Beyond Cultural Capital: Toward a Theory of Symbolic Domination, in D. Robbins (ed.), *Pierre Bourdieu,* London, Thousand Oaks, CA and New Delhi, SAGE Publications, vol. 1, pp. 278–301.

McAleer, K. (1994) *Dueling: The Cult of Honor in Fin-de-Siècle Germany,* Princeton, NJ, Princeton University Press.

McClelland, C. E. (1991) *The German Experience of Professionalization. Modern Learned Professions and Their Organisations from the Early Nineteenth Century to the Hitler Era,* Cambridge, Cambridge University Press.

McClelland, C. E. (1997) Modern German Doctors: A Failure of Professionalization?, in M. Berg and G. Cocks (eds), *Medicine and Modernity: Public Health and Medical Care in Nineteenth- and Twentieth-Century Germany,* Cambridge, Cambridge University Press, pp. 81–97.

McCullough, L. B. (ed.) (1998) *John Gregory's Writings on Medical Ethics and Philosophy of Medicine,* Dordrecht and Boston, MA, Kluwer Academic Publishers.

Mohr, J. C. (1996) *Doctors and the Law: Medical Jurisprudence in Nineteenth-Century America,* Baltimore, MD, Johns Hopkins University Press.

Morrice, A. A. G. (2002a) 'Honour and Interests': Medical Ethics and the British Medical Association, in A.-H. Maehle and J. Geyer-Kordesch (eds),

Historical and Philosophical Perspectives on Biomedical Ethics. From Paternalism to Autonomy?, Aldershot and Burlington, VT, Ashgate, pp. 11–35.

Morrice, A. A. G. (2002b) 'Should the Doctor Tell?' Medical Secrecy in Early Twentieth-Century Britain', in S. Sturdy (ed.), *Medicine, Health and the Public Sphere in Britain, 1600–2000*, London and New York, Routledge, pp. 61–82.

Moscucci, O. (1993) *The Science of Woman. Gynaecology and Gender in England 1800–1929*, Cambridge, Cambridge University Press.

Müller, C. (2004) *Verbrechensbekämpfung im Anstaltsstaat: Psychiatrie, Kriminologie und Strafrechtsreform in Deutschland 1871–1933*, Göttingen, Vandenhoeck & Ruprecht.

Noack, T. (2004) *Eingriffe in das Selbstbestimmungsrecht des Patienten: Juristische Entscheidungen, Politik und ärztliche Positionen 1890–1960*, Frankfurt am Main, Mabuse-Verlag.

Nolte, K. (2007) Zeitalter des ärztlichen Paternalismus? – Überlegungen zu Aufklärung und Einwilligung von Patienten im 19. Jahrhundert, *Medizin, Gesellschaft und Geschichte* 25, pp. 59–89.

Nolte, K. (2008) 'Telling the Painful Truth' – Nurses and Physicians in the Nineteenth Century, *Nursing History Review* 16, pp. 115–134.

Nutton, V. (1993) Beyond the Hippocratic Oath, in A. Wear, J. Geyer-Kordesch and R. French (eds), *Doctors and Ethics: The Earlier Historical Setting of Professional Ethics*, Amsterdam and Atlanta, GA, Rodopi, pp. 10–37.

Nye, R. A. (1995) Honor Codes and Medical Ethics in Modern France, *Bulletin of the History of Medicine* 69, pp. 91–111.

Nye, R. A. (1997) Medicine and Science as Masculine 'Fields of Honor', *Osiris* 12, pp. 60–79.

Ostrow, J. M. (2000) Culture as a Fundamental Dimension of Experience: A Discussion of Pierre Bourdieu's Theory of Human Habitus, in D. Robbins (ed.), *Pierre Bourdieu*, London, Thousand Oaks, CA and New Delhi, SAGE Publications, vol. 1, pp. 302–322.

Pagel, W. (1951) Julius Pagel and the Significance of Medical History for Medicine, *Bulletin of the History of Medicine* 25, pp. 207–225.

Pernick, M. (1982) The Patient's Role in Medical Decisionmaking: A Social History of Informed Consent in Medical Therapy, in President's Commission for the Study of Ethical Problems in Medicine and Biomedical and Behavioral Research (ed.), *Making Health Care Decisions: The Ethical and Legal Implications of Informed Consent in the Patient-Practitioner Relationship*, Washington, DC, vol. 3, pp. 1–35.

Pranghofer, S. and Maehle, A.-H. (2006) Limits of Professional Secrecy: Medical Confidentiality in England and Germany in the Nineteenth and Early Twentieth Centuries, *Interdisciplinary Science Reviews* 31, pp. 231–244.

Proctor, R. (1988) *Racial Hygiene: Medicine under the Nazis*, Cambridge, MA and London, Harvard University Press.

Prüll, C.-R. and Sinn, M. (2002) Problems of Consent to Surgical Procedures and Autopsies in Twentieth Century Germany, in A.-H. Maehle and J. Geyer-Kordesch (eds), *Historical and Philosophical Perspectives on Biomedical Ethics*.

From Paternalism to Autonomy?, Aldershot and Burlington, VT, Ashgate, pp. 73–93.

Putzke, S. (2003) *Die Strafbarkeit der Abtreibung in der Kaiserzeit und in der Weimarer Zeit. Eine Analyse der Reformdiskussion und der Straftatbestände in den Reformentwürfen*, Berlin, Berliner Wissenschafts-Verlag.

Rabi, B. (2002) *Ärztliche Ethik – Eine Frage der Ehre? Die Prozesse und Urteile der ärztlichen Ehrengerichtshöfe in Preußen und Sachsen 1918–1933*, Frankfurt/Main, Peter Lang.

Regin, C. (1995) *Selbsthilfe und Gesundheitspolitik. Die Naturheilbewegung im Kaiserreich (1889–1914)*, Stuttgart, Franz Steiner Verlag.

Riegger, T. (2007) *Die historische Entwicklung der Arzthaftung*, Jur. Diss. Regensburg.

Riha, O. (1995) Die Krankenakten der Chirurgischen Universitätsklinik Göttingen als Quelle der Medizingeschichte (1912–1950). Möglichkeiten und Grenzen einer Methode, *Würzburger medizinhistorische Mitteilungen* 13, pp. 5–15.

Ritzmann, I. (1999) Der Verhaltenskodex des 'Savoir faire' als Deckmantel ärztlicher Hilflosigkeit, *Gesnerus* 56, pp. 197–219.

Robbins, D. (2000) *Bourdieu and Culture*, London, Thousand Oaks, CA and New Delhi, SAGE Publications.

Rothschuh, K. E. (1983) Das Buch 'Der Arzt' (1906) stammt nicht von Ernst Schweninger!, *Medizinhistorisches Journal* 18, pp. 137–144.

Rothschuh, K. E. (1984) Ernst Schweninger (1850–1924). Zu seinem Leben und Wirken. Ergänzungen, Korrekturen, *Medizinhistorisches Journal* 19, pp. 250–258.

Rüther, M. (1997) Ärztliches Standeswesen im Nationalsozialismus 1933–1945, in R. Jütte (ed.), *Geschichte der deutschen Ärzteschaft. Organisierte Berufs- und Gesundheitspolitik im 19. und 20. Jahrhundert*, Cologne, Deutscher Ärzte-Verlag, pp. 143–193.

Sauerteig, L. (1999) *Krankheit, Sexualität, Gesellschaft. Geschlechtskrankheiten und Gesundheitspolitik in Deutschland im 19. und frühen 20. Jahrhundert*, Stuttgart, Franz Steiner Verlag.

Sauerteig, L. (2000) Ethische Richtlinien, Patientenrechte und ärztliches Verhalten bei der Arzneimittelerprobung (1892–1931), *Medizinhistorisches Journal* 35, pp. 303–334.

Sauerteig, L. (2001) 'The Fatherland Is in Danger, Save the Fatherland!' Venereal Disease, Sexuality and Gender in Imperial and Weimar Germany, in R. Davidson and L. A. Hall (eds), *Sex, Sin and Suffering. Venereal Disease and European Society since 1870*, London and New York, Routledge, pp. 76–92.

Schivelbusch, W. (2003) *The Culture of Defeat. On National Trauma, Mourning, and Recovery*, translated by J. Chase, New York, Metropolitan Books.

Schmiedebach, H.-P. (1989) Der wahre Arzt und das Wunder der Heilkunde. Erwin Lieks ärztlich-heilkundliche Ganzheitsideen, *Argument-Sonderband (Der ganze Mensch und die Medizin)* 162, pp. 33–53.

Schmiedebach, H.-P. (1996) Eine 'antipsychiatrische Bewegung' um die Jahrhundertwende, in M. Dinges (ed.), *Medizinkritische Bewegungen im*

Deutschen Reich (ca. 1870 – ca. 1933), Stuttgart, Franz Steiner Verlag, pp. 127–159.

Schmiedebach, H.-P. (2001) Medizinethik und 'Rationalisierung' im Umfeld des Ersten Weltkrieges, in A. Frewer and J. N. Neumann (eds), *Medizingeschichte und Medizinethik: Kontroversen und Begründungsansätze 1900–1950*, Frankfurt and New York, Campus Verlag, pp. 57–75.

Schmuhl, H.-W. (1992) *Rassenhygiene, Nationalsozialismus, Euthanasie: Von der Verhütung zur Vernichtung 'lebensunwerten Lebens', 1890–1945*, 2nd edn, Göttingen, Vandenhoeck & Ruprecht.

Schomerus, G. (2001) *Ein Ideal und sein Nutzen. Ärztliche Ethik in England und Deutschland 1902–1933*, Frankfurt/Main, Peter Lang.

Schultz, J. H. (1986) *Albert Molls Ärztliche Ethik*, MD thesis, Zurich, Juris-Verlag.

Seidler, E. (1993) Das 19. Jahrhundert. Zur Vorgeschichte des Paragraphen 218, in R. Jütte (ed.), *Geschichte der Abtreibung. Von der Antike bis zur Gegenwart*, Munich, C. H. Beck, pp. 120–139.

Siegrist, H. (1990) Public Office or Free Profession? German Attorneys in the Nineteenth and Early Twentieth Centuries, in G. Cocks and K. H. Jarausch (eds), *German Professions, 1800–1950*, New York and Oxford, Oxford University Press, pp. 46–65.

Sinn, M. (2001) *Einwilligung und Aufklärung vor operativen Eingriffen in Deutschland 1894–1945: '... der Kranke bekommt davon, soviel er nötig hat.'*, MD thesis, Albert-Ludwigs-Universität Freiburg im Breisgau.

Smith, R. (2007) *Being Human. Historical Knowledge and the Creation of Human Nature*, Manchester and New York, Manchester University Press.

Smith, R. G. (1994) *Medical Discipline. The Professional Conduct Jurisdiction of the General Medical Council, 1858–1990*, Oxford, Clarendon Press.

Spree, R. (1989) Kurpfuscherei-Bekämpfung und ihre sozialen Funktionen während des 19. und zu Beginn des 20. Jahrhunderts, in A. Labisch and R. Spree (eds), *Medizinische Deutungsmacht im sozialen Wandel des 19. und 20. Jahrhunderts*, Bonn, Psychiatrie-Verlag, pp. 103–121.

Steinacher, G. M. P. (1959) *Das Leben und Wirken Ottomar Rosenbachs unter besonderer Berücksichtigung der Energetik*, Dent. MD thesis, LMU Munich.

Stolberg, M. (1998) Heilkundige: Professionalisierung und Medikalisierung, in N. Paul and T. Schlich (eds), *Medizingeschichte: Aufgaben, Probleme, Perspektiven*, Frankfurt/Main and New York, Campus Verlag, pp. 69–86.

Stolberg, M. (2007a) 'Cura palliativa'. Begriff und Diskussion der palliativen Krankheitsbehandlung in der vormodernen Medizin (ca. 1500–1850), *Medizinhistorisches Journal* 42, pp. 7–29.

Stolberg, M. (2007b) Active Euthanasia in Pre-Modern Society, 1500–1800: Learned Debates and Popular Practices, *Social History of Medicine* 20, pp. 205–221.

Stolberg, M. (2008) Two German Pioneers of Euthanasia around 1800, *Hastings Center Report* 38, no. 3, pp. 19–22.

Strätling, M. (1998) *Die Begründung der neuzeitlichen Medizinethik in Praxis, Lehre und Forschung: John Gregory (1724–1773) und seine Lectures on the Duties and Qualifications of a Physician*, Frankfurt/Main, Lang.

Tannenbaum, R. J. (1994) Earnestness, Temperance, Industry: The Definition and Uses of Professional Character Among Nineteenth-Century American Physicians, *Journal of the History of Medicine and Allied Sciences* 49, pp. 251–283.

Thomsen, P. (1996) *Ärzte auf dem Weg ins 'Dritte Reich'. Studien zur Arbeitsmarktsituation, zum Selbstverständnis und zur Standespolitik der Ärzteschaft gegenüber der staatlichen Sozialversicherung während der Weimarer Republik*, Husum, Matthiesen Verlag.

Tod, R. (1979) Le secret médical au XIXe sciècle dans les textes médicaux et juridiques, *La Nouvelle Presse Médicale* 8, pp. 2695–2697.

Tröhler, U. (1993) Surgery (Modern), in W. F. Bynum and R. Porter (eds), *Companion Encyclopedia of the History of Medicine*, London and New York, Routledge, vol. 2, pp. 984–1028.

Tröhler, U. and Maehle, A.-H. (1990) Anti-vivisection in Nineteenth-Century Germany and Switzerland: Motives and Methods, in N. A. Rupke (ed.), *Vivisection in Historical Perspective*, London and New York, Routledge, pp. 149–187.

Tüllmann, G. (1938) *Eduard Graf ein deutscher Arzt und Ärzteführer*, MD thesis Düsseldorf 1937, Ochsenfurt am Main, Fritz & Rappert.

Usborne, C. (1992) *The Politics of the Body in Weimar Germany: Women's Reproductive Rights and Duties*, London, Macmillan.

Usborne, C. (2007) *Cultures of Abortion in Weimar Germany*, New York and Oxford, Berghahn Books.

Villey, R. (1986) *Histoire du Secret Médical*, Paris, Seghers.

Vogt, G. (1998) *Ärztliche Selbstverwaltung im Wandel. Eine historische Dokumentation am Beispiel der Ärztekammer Nordrhein*, Cologne, Deutscher Ärzte-Verlag.

Vogt, L. (1997) *Zur Logik der Ehre in der Gegenwartsgesellschaft: Differenzierung, Macht, Integration*, Frankfurt/Main, Suhrkamp.

Waddington, I. (1984) *The Medical Profession in the Industrial Revolution*, Dublin, Gill and Macmillan.

Wahrig, B. and Sohn, W. (eds) (2003) *Zwischen Aufklärung, Policey und Verwaltung. Zur Genese des Medizinalwesens 1750–1850*, Wiesbaden, Harrassowitz Verlag.

Wear, A., Geyer-Kordesch, J. and French, R. (eds) (1993) *Doctors and Ethics: The Earlier Historical Setting of Professional Ethics*, Amsterdam and Atlanta, GA, Rodopi.

Weindling, P. (1989) *Health, Race and German Politics between National Unification and Nazism, 1870–1945*, Cambridge, Cambridge University Press.

Weindling, P. (1990) Bourgeois Values, Doctors and the State: The Professionalization of Medicine in Germany 1848–1933, in D. Blackbourn and R. J. Evans (eds), *The German Bourgeoisie: Essays on the Social History of the German Middle Class from the Late Eighteenth to the Early Twentieth Century*, London, Routledge, pp. 198–223.

Weindling, P. (2004) *Nazi Medicine and the Nuremberg Trials. From Medical War Crimes to Informed Consent*, Basingstoke and New York, Palgrave Macmillan.

Wetzell, R. F. (1996) The Medicalization of Criminal Law Reform in Imperial Germany, in N. Finzsch and R. Jütte (eds), *Institutions of Confinement. Hospitals, Asylums, and Prisons in Western Europe and North America, 1500–1950*, Cambridge, Cambridge University Press, pp. 275–283.

Wiesemann, C. (1997) Das Recht auf Selbstbestimmung und das Arzt-Patient-Verhältnis aus sozialgeschichtlicher Perspektive, in R. Toellner and U. Wiesing (eds), *Geschichte und Ethik in der Medizin. Von den Schwierigkeiten einer Kooperation*, Stuttgart, Fischer, pp. 67–90.

Wiesing, U. (1996) Die Persönlichkeit des Arztes und das geschichtliche Selbstverständnis der Medizin. Zur Medizintheorie von Ernst Schweninger, Georg Honigmann und Erwin Liek, *Medizinhistorisches Journal* 31, pp. 181–208.

Wimmer, W. (1992) Die Pharmazeutische Industrie als 'ernsthafte Industrie': Die Auseinandersetzung um die Laienwerbung im Kaiserreich, *Medizin, Gesellschaft und Geschichte* 11, pp. 73–86.

Wimmer, W. (1994) *'Wir haben fast immer was Neues': Gesundheitswesen und Innovationen der Pharma-Industrie in Deutschland, 1880–1935*, Berlin, Duncker & Humblot.

Winau, R. (1996) Medizin und Menschenversuch. Zur Geschichte des 'informed consent', in C. Wiesemann and A. Frewer (eds), *Medizin und Ethik im Zeichen von Auschwitz: 50 Jahre Nürnberger Ärzteprozeß*, Erlangen and Jena, Palm & Enke, pp. 13–29.

Winkelmann, O. (1964) Anfänge und Aufgaben des Ehrengerichts der Ärztekammer Berlin-Brandenburg vor 1914, *Berliner Medizin* 15, pp. 323–325.

Winkelmann, O. (1992) Albert Moll als Sexualwissenschaftler und Sexualpolitiker, in R. Gindorf and E. J. Haeberle (eds), *Sexualwissenschaft und Sexualpolitik: Spannungsverhältnisse in Europa, Amerika und Asien*, Berlin and New York, Walter de Gruyter, pp. 65–71.

Winkelmann, O. (1996) Der vergessene Albert Moll (1862–1939) und sein 'Leben als Arzt der Seele', in N. Goldenbogen, S. Hahn, C.-P. Heidel and A. Scholz (eds), *Medizinische Wissenschaften und Judentum*, Dresden, Verein für regionale Politik und Geschichte, pp. 46–52.

Wolff, E. (1996) Medizinkritik der Impfgegner im Spannungsfeld zwischen Lebenswelt- und Wissenschaftsorientierung, in M. Dinges (ed.), *Medizinkritische Bewegungen im Deutschen Reich (ca. 1870 – ca. 1933)*, Stuttgart, Franz Steiner Verlag, pp. 79–108.

Wolff, E. (1997) Mehr als nur materielle Interessen: Die organisierte Ärzteschaft im Ersten Weltkrieg und in der Weimarer Republik 1914–1933, in R. Jütte (ed.), *Geschichte der deutschen Ärzteschaft. Organisierte Berufs- und Gesundheitspolitik im 19. und 20. Jahrhundert*, Cologne, Deutscher Ärzte-Verlag, pp. 97–142.

Woycke, J. (1988) *Birth Control in Germany 1871–1933*, London and New York, Routledge.

Woycke, J. (1992) Patent Medicines in Imperial Germany, *Canadian Bulletin of Medical History* 9, pp. 41–56.

Index

Note: Page numbers in *italics* denote pages containing tables.